# Exploring Religious Dimensions in AI and Humanity

# Exploring Religious Dimensions in AI and Humanity

Aristarchos-Vasileios Gkrekas
Nicholas Kokkinos

LEXINGTON BOOKS
*Lanham • Boulder • New York • London*

Published by Lexington Books
An imprint of The Rowman & Littlefield Publishing Group, Inc.
4501 Forbes Boulevard, Suite 200, Lanham, Maryland 20706
www.rowman.com

86-90 Paul Street, London EC2A 4NE, United Kingdom



British Library Cataloguing in Publication Information Available

**Library of Congress Cataloging-in-Publication Data**

Names: Gkrekas, Aristarchos Vasileios, 1968– author. | Kokkinos, Nicholas, 1969– author.
Title: Exploring religious dimensions in AI and humanity / Nicholas Kokkinos and Aristarchos-Vasileios Gkrekas.
Description: Lanham : Lexington Books, [2024] | Includes bibliographical references and index. | Summary: "This book explores how the phenomenal advancements of artificial intelligence can reshape our spiritual journeys and influence established ethical frameworks. It is an essential read for those intrigued by the evolving dialogue between technology and spirituality and is enriched by references to centuries-old texts"—Provided by publisher.
Identifiers: LCCN 2024010765 (print) | LCCN 2024010766 (ebook) | ISBN 9781666964134 (cloth) | ISBN 9781666964141 (epub)
Subjects: LCSH: Artificial intelligence—Religious aspects. | Artificial intelligence—Moral and ethical aspects. | Transhumanism—Religious aspects. | Transhumanism—Moral and ethical aspects.
Classification: LCC Q334.7 .K664 2024 (print) | LCC Q334.7 (ebook) | DDC 201/.60063—dc23/eng/20240530
LC record available at https://lccn.loc.gov/2024010765
LC ebook record available at https://lccn.loc.gov/2024010766

™ The paper used in this publication meets the minimum requirements of American National Standard for Information Sciences—Permanence of Paper for Printed Library Materials, ANSI/NISO Z39.48-1992.

# Contents

# Preface

Artificial intelligence represents a unique technological advancement in the Modern Age. While its applications aim to facilitate day-to-day life, they also give birth to immense challenges that raise important questions.

Since artificial intelligence interacts more with humans, global literature explores the ethical implications of its uses and how it influences human perception in a new emerging global environment.

Our goal is first to explore these influences and second to provide the ethical answers to these implications as seen from the angle of religion and ethical point of view. In that light, the book offers numerous references from ancient religious texts, especially from Christianity, to give answers or to demystify the ethical puzzles that arise from this new technology.

To create a bridge between religion and artificial intelligence, it was imperative to consider religious and moral frameworks that have existed for almost 2,000 years and how they apply in modern society. In addition, research has been performed on if these innovative systems align with and respect human dignity, liberty, and the broader ethical principles within a wider religious context. It is a pathway that doesn't restrict but, in contrast, enriches artificial intelligence.

These problems posed by the interaction of AI and humans, as well as new approaches to the concept of humanity in modern reality, puzzled both authors and inspired Dr. Aristarchos Gkrekas to attend and complete the Executive Programme at the Harvard Kennedy School, which is only open to Harvard University graduates. The program, "Artificial Intelligence: Exploring Policy, Technology, and Governance," concluded in May 2019. It was a one-of-a-kind encounter attended by academics and top specialists with current knowledge of artificial intelligence progress.

We feel morally bound to offer our deepest gratitude to His Beatitude Ieronymos II, Archbishop of Athens and All Greece, for his fatherly support in achieving our goal.

# List of Abbreviations

| | |
|---|---|
| AAI | apocalyptic artificial intelligence |
| AGI | artificial general intelligence |
| AI | artificial intelligence |
| ANN | artificial neural networks |
| BBC | British Broadcasting Corporation |
| CBS | Columbia Broadcasting System |
| EU | European Union |
| G20 | Group of Twenty |
| GCF | Greek Church Fathers |
| GDPR | General Data Protection Regulation |
| HCAI | human-centered artificial intelligence |
| HGCOC | Holy and Great Council of the Orthodox Church |
| IEEE | Institute of Electrical and Electronics Engineers |
| LGCFTA | Literature of the Greek Church Fathers and Theological Authors |
| ML | machine learning |
| NBC | National Broadcasting Company |
| NLP | Natural Language Processing |
| PG | Patrologia Graeca, J.-P. Migne |
| SC | Sources Chrétiennes |
| SYSTUA | Scientific Yearbook of the School of Theology of the University of Athens |
| SYSTUT | Scientific Yearbook of the School of Theology of the University of Thessaloniki |
| UDHR | Universal Declaration of Human Rights |
| UK | United Kingdom |
| UN | United Nations |
| UNI | Union Network International |
| US | United States |
| WOTF | Way of the Future |

# Introduction

Technological progress, including artificial intelligence (AI), has grown impressively in recent years. The emerging potential is unparalleled in an increasingly globalized world regarding communication, instantaneous distribution of news, social media use, and socioeconomic data.

With the above mentioned information in mind, the escalating volume of data and the demand for efficient processing underscore the necessity of artificial intelligence, which offers significant benefits in resource management and data analysis, alongside presenting unique ethical challenges. Artificial intelligence covers a wide range of everyday applications to assist citizens, governments, the economy, and organizations, providing outstanding services with an exceptional economy of resources, time, and rapid data analysis. This burst of progress and usability of artificial intelligence makes it indispensable, along with the ethical dilemmas that emerge upon its application.

In part I, there is an analysis of the term "artificial intelligence," its use, and its applications. Following this, the concept of human-centered artificial intelligence and the implications for the human-centered ethics of AI are discussed and analyzed. In human-centered artificial intelligence, a reference is made to the codes of ethical conduct concerning human values, human rights, and its governance applications.

Of research significance is the exploration of the theological perspective, which discusses the use and implications of AI in general and its approach to salvation and immortality. In the context of this approach to AI, as documented in the literature, the trend of "apocalyptic" AI is explored, and the tendency to acknowledge AI as a religion or a church with the respective ceremonial uses of the AI applications. Part I concludes with a skeptical view of the malicious uses from both pragmatic and theological perspectives.

Part II first examines the theological approach of artificial intelligence in relation to humans. However, it is essential to elaborate on the Orthodox theological teachings on humanity according to the Eastern Christian Orthodox theology for an adequate understanding of approaching humans as a psychosomatic unity created by God as the pinnacle of divine creation. In addition to analyzing the relationship between the soul and the body of the human, the ecclesiastical literature and theology are presented, and the meaning of the "deified" human as the "new human" in Jesus Christ is further discussed.

Subsequently, it is considered necessary to describe the true character of the Orthodox Church as it has been defined and experienced by ecclesiastical tradition and theology. At the same time, the ecclesiological dimension of humans as God's creation is presented through the Christological approach. The importance of the devotional and sacramental aspects of the body of the Church, the relevance of ethics, and the importance of Love are also developed. In the context of Orthodox anthropology, the book's focal point is the soteriological approach in correlation with the Orthodox "Revelation" theology, as well as the "theory" as a medium for personal revelation.

Based on the points above, it is necessary to reflect on humans from a theological approach so that the applications and development of AI will ensure religious self-consciousness and freedom against exaggerated viewpoints on the consequences of AI in our everyday lives. In this instance, the Orthodox Church, on account of its traditions, provides reliable anthropological and theological criteria for human self-consciousness, relationship with God, and the proper use of human advancements in technology.

# PART I

# *Chapter 1*

# Artificial Intelligence

The first step towards understanding this topic requires defining what intelligence is. The term "intelligence" involves the processes associated with memory, imagination, associative thinking, perception, intellect, and attentiveness, regarded as a set of human cognitive abilities. Thus, the ability to adapt to new situations can be developed[1] through a comparative process that associates commonalities, differences, and convergences. In addition, intelligence refers to the mental functions used to develop coping strategies for new circumstances and problem-solving skills by utilizing past experiences.

The seven main types of intelligence are as follows: Linguistic Intelligence, involving the effective use of language, Logical-Mathematical Intelligence, Spatial Intelligence, Musical Intelligence, Bodily-Kinesthetic Intelligence, Interpersonal Intelligence, and Intra-personal Intelligence.

For some, AI could be regarded as a new science, but its development occurs in conjunction with other fields of science, such as engineering, psychology, linguistics, physics, and mathematics. Scientists consider Aristotle to be one of the forerunners of AI on a theoretical level in terms of the evolutionary process of logical thinking, i.e., the process of irrefutable and linear reasoning. Models of logical reasoning expressions lead to sound conclusions based on correct assumptions as they explore the functionality of syllogisms within the field of Logic.[2]

Different aspects and suggestions have been presented to define what AI is, but given the rapidly evolving nature of this technology, the definition is constantly changing and diversifying.[3] There are some definitions provided by scientists, institutes, or international organizations, such as the IEEE's definition of "Autonomous Intelligent Systems."[4]

Certain aspects of an AI's definition are also found in other texts. For instance, some documents portray artificial intelligence as a set of systems or systems that autonomously engage in actions to accomplish predefined goals, and others include that these actions are generally tasks that would otherwise require human intelligence.[5] Definitions from a descriptive approach will be given below.

More recently, given the need to establish a common European legal framework with aligned definitions and common ethical principles, the European Union has defined AI in a way that includes its use for defense purposes. Subsequently, the following definitions have been adopted by the designated Commission: "Artificial intelligence (AI) refers to systems that display intelligent behavior by analyzing their environment and taking actions—with some degree of autonomy—to achieve specific goals. AI-based systems can be purely software-based, acting in the virtual world (e.g., voice assistants, image analysis software, search engines, speech, and face recognition systems) or AI can be embedded in hardware devices (e.g., advanced robots, autonomous cars, drones or Internet of Things applications)."[6]

The origins of artificial intelligence are rooted back in the 1950s when scientists and philosophers began exploring the concept of creating machines smart enough to perform tasks that ordinarily require human intelligence. An early focus of such research was the development of algorithms capable of analyzing data to perform specified tasks, such as playing chess or translating languages. As AI technology has advanced, its capabilities have greatly expanded, and it is now used in a vast array of applications, such as self-driving vehicles, personal assistants, and medical diagnoses.

According to Gardner,[7] AI innovations fall into four distinctive categories:

- Data-centric AI, where the focus is on enriching data to better train the algorithms;
- Model-centric AI, where the intention is towards better outputs and knowledge representation;
- Application-centric AI to improve the decision-making process; and, finally,
- Human-centric AI, where that new technology calculates ethical risks as an outcome of replacing human decisions.

According to the same study, some of these innovations are available now and some will hit us in two to five years from now.

Innovation uses the intersection of new and existing ideas to produce advancements, and historically, technological leaps invigorate such a change. A phenomenon that is always approached with a critical eye. Only a small percentage of people adopt innovation directly and others follow over time.[8] For instance, Henry Ford's famous quote about cars, "If I had asked people what they wanted, they would have said faster horses," shows that the public needs an additional parameter to digest that change. And that parameter is time.

Nevertheless. the intersection of artificial intelligence and religion raises numerous concerns as does any other innovation. Questions arise regarding the nature of consciousness or the potential for AI to possess moral agency. It also presents practical issues for the future of both disciplines, such as the potential for AI to alter religious beliefs.

The history of artificial intelligence in philosophy and computer science is extensive and complex. Ancient Greek mythology and the legend of Pygmalion,[9] who created Galatea, a statue of a woman that came to life, provides the origins of the concept of making intelligent machinery.

Alan Turing, a British mathematician known for his work on the Enigma machine during World War II and his contributions to developing the modern computer, was one of the earliest pioneers in AI. In 1950, Turing published a paper entitled "Computing Machinery and Intelligence,"[10] He proposed the "Turing Test" to determine whether a machine can exhibit cognitive behavior equivalent to or indistinguishable from a human's.

In the following decades, AI research focused on developing algorithms and computer programs to perform tasks such as playing chess or translating languages. This type of artificial intelligence is considered narrow or weak, as it is designed to perform a specific task rather than manifest general intelligence.

In the 1960s and 1970s, AI researchers began investigating the concept of artificial neural networks, designed to process and analyze large amounts of data and modelled after the human brain's neural networks. In the 1980s and 1990s, interest in artificial intelligence was revived, fuelled partly by computer hardware and software advances. Expert systems and computer programs using a knowledge database to make decisions in a specific domain were developed during this period. Various industries, including medicine and finance, utilized expert systems to assist with decision-making duties.

AI has continued to evolve and expand its capabilities in the twenty-first century. For instance, the development of machine learning (ML), a subfield of artificial intelligence that employs algorithms and statistical models to enable computers to learn and enhance their performance over time, has resulted in significant advancements in the field. Nowadays, applications use machine learning algorithms, including image and speech recognition, language translation, and autonomous vehicles.

AI systems can either use symbolic rules or learn a numerical model and can also adapt their behavior by analyzing how the environment is affected by previous actions.[11]

In recent definitions, the term "imitation of intelligent human behavior" has frequently been encountered, as the AI technology community tries to mimic intelligent behavior through computer programs. This is not always an easy task because a computer program must be able to do many different things to be considered intelligent. Rather than giving a broad definition when referring to AI, we can narrow it down to considering AI as a technology of artificially intelligent systems. In other words, this is not a universal definition but rather a reference on a functional level since AI approaches human's mental functioning, particularly intelligence and intellect, in a simulative and generative way. For this reason, considering the multiple functional orientations of AI systems, corresponding needs for classification into groups and models arise, which—as mentioned above—take the form of definitions associated with the attempt to categorize artificially intelligent systems. A more accurate approach is to classify most of these systems into the following four categories: systems that think like humans, systems that act like humans, systems that think rationally, and systems that act rationally.[12]

And here is an interesting point: One of the primary ways AI challenges our comprehension of human nature is by redefining intelligence. Intelligence has traditionally been considered a defining characteristic of humans, distinguishing us from other animals and setting us apart as rational beings. Nevertheless, AI has already demonstrated that machines can replicate or even surpass many cognitive tasks once believed to be exclusive to humans. Computing systems have demonstrated extraordinary problem-solving and decision-making abilities, from playing complex games to diagnosing medical conditions and operating autonomous vehicles.

The advent of artificial intelligence compels us to consider whether intelligence is innate to humans or merely the result of intricate information processing. This question has far-reaching implications for our comprehension of human nature because it challenges the long-held belief that our intelligence distinguishes us from other beings. Moreover, the development of AI raises concerns regarding the uniqueness of human creativity and innovation, as a simple user with no background can now produce art, music, and literature that is frequently indistinguishable from human-created works. Solutions like Adobe Firefly, Midjourney.com, and DALL·E 3 feature unimaginable capabilities even five years ago.

Generative algorithms are currently employed in several fields, such as painting, music composition, and storytelling. These applications have the potential to challenge the distinction between human creativity and content generated by machines, prompting thought-provoking inquiries regarding the concepts of authorship and originality. The possibility of AI in creative applications holds promise, but it also presents a multifaceted array of implications yet to be comprehensively examined.

Another domain affects our comprehension of human nature and the self by challenging conventional beliefs regarding emotion and consciousness. These systems can now, for example, recognize and respond to human emotions and elicit emotional responses. This raises queries regarding the nature of emotion and whether it is a characteristic inherent to humans or the result of complex information processing, like intelligence. That question was obvious in successful television series (e.g., *Star Trek*) when fictional characters like Spock excelled in logic but lacked human emotions. Their logical thinking does not permit them to be influenced by feelings but by facts only.

One of the most difficult tasks in AI is creating systems that can recognize and respond to human emotions. The ability to sense, interpret, and manage emotions is referred to as emotional intelligence. It is a necessary talent for humans to engage with one another effectively.

Building AI systems that recognize and respond to human emotions offers numerous advantages. They may, for example, be used to create more successful, sophisticated customer-service systems, educational programs, and healthcare treatments. They could also be utilized to create more engaging and immersive virtual worlds and video games.

In a variety of ways, AI has the potential to augment human emotional intelligence. AI systems, for example, can be used to:

1. Analyze enormous amounts of data to find patterns and trends people need help discerning. These data can then be used to make more informed decisions and devise more successful tactics.
2. Provide individuals with personalized feedback and support. AI-powered educational programs, for example, may customize their training to each student's unique needs.
3. Generate immersive experiences that can assist individuals in developing their learning skills. Gamification is a powerful tool that in combination with artificial intelligence can provide highly engaging experiences.

Furthermore, the advancement of artificial intelligence has implications for our comprehension of the self and individual identity. Technology and the internet influence our decision-making, social interactions, and even our sense of self as they become more integrated into our daily lives. For example, AI-powered recommendation systems and social media algorithms can influence our preferences, beliefs, and values, raising questions about our decisions' autonomy and the extent to which external forces shape our identities. Moreover, as AI technology becomes more capable of simulating human-like behavior, it challenges our presumptions regarding the nature of personhood and what it means to be a distinct individual.[13]

AI has the potential to not only revolutionize our daily lives and societal structures but also to challenge long-held philosophical beliefs, such as the concept of free will. An idea that humans can make decisions and act on their own volition rather than being determined by external factors or forces. Here, the following question comes to our mind: Does artificial intelligence have the capacity to exhibit free will? An effective method for answering this question is examining the nature of algorithmic decision-making. Before choosing the most suitable action, AI decision-making often utilizes sophisticated algorithms assessing available data—usually tons of them—and calculating among a plethora of potential outcomes. It is like the human decision-making process, which involves evaluating possibilities and analyzing outcomes. An interesting interplay of biology, culture, and individual experience that influences human decision-making.

The function of consciousness in free will is a further crucial consideration. Many believe that consciousness is necessary for free will because it enables us to recognize our options and make decisions based on our values and beliefs. There is no scientific consensus on whether AI systems can be conscious and, if so, what form consciousness may take in an artificial system.[14]

AI is a synthesis of tools, systems, methodologies, and models which, via AI, are composed and simultaneously subjected to specialization in four different scientific areas related to the corresponding Cognitive sciences, namely Cybernetics, Philosophy, Psychology, and Reasoning. They are also all connected with the Mathematical Sciences, the Proof Theory and Model Theory, Mathematical Logic and Operational Research, as well as with Computer Science, Intelligent Information Systems, and Control Theory combined with Automatics and Biotechnology.[15]

AI also incorporates various techniques and approaches, such as machine learning and machine reasoning, which includes programming, design, search optimization, knowledge representation, and reasoning, as well as robotics, which contains perception, control, actuators, and sensors, and the incorporation of techniques in cyber-physical systems.[16]

## APPLICATIONS OF AI

The dynamic progress of AI reinforces its importance in combination with its usability due to the increased capacity to store and process data. It is undeniably a technological revolution with the potential for processing and interconnecting the Internet with the Big Data environment and Learning Machines. AI is taking advantage of the existing potential of technology and the advancement of technologies, quickly creating opportunities and providing solutions to issues related to everyday life, the environment, transportation, health improvement, and even governance.[17]

AI, across its multiple applications, can comprehend the focus areas in a short time and make optimal decisions. Moreover, artificial intelligence could be considered a computer system that can function cognitively with exceptional learning adaptability in the most efficient way possible. For example, they interpret their environment through fast processing of data and signal streams such as Natural Language and Processing Computer Vision technologies that enable computer systems to "understand" speech, video, and images to react in the most effective way possible.[18]

AI is divided into two categories in terms of approach towards its development and application: symbolic or classical and associative or computational AI. Classical intelligence is predicated on understanding mental processes with the potential of simulating human intelligence. This process of understanding and simulation employs systems and algorithms with the assistance of symbols. Computational intelligence mimics biological processes, such as how the human brain works or the evolutionary process of species.[19]

Modern AI applications apply to Cognitive Science and are closely related to learning, natural interface applications, and robotics applications. Cognitive Science applications include applications of expert systems, fuzzy logic systems, adaptive intelligent learning environments, intelligent agents, and applications supported by machine learning systems, such as genetic algorithms and neural networks.[20]

Natural interface applications refer to those applications that have and continue to develop the ability to communicate in a natural language through voice recognition, as well as interface environments that have the capacity to feature multiple sensors, creating the perception of ambient intelligence, as demonstrated by virtual reality environments. Robotic applications involve machines capable of visual perception, dexterity, movement of parts, touch, and spatial navigation.[21]

The applications of AI systems include the Theorem Proving, which results from the strong connection between mathematics and artificial intelligence. Natural Language Processing with the algorithmic input for adapting syntactic, semantic, and pragmatic analyses to a wide range of applications.[22] In addition, Artificial-Computer Vision with typical applications based on AI technologies, such as Computer Vision when algorithms acquire vision.

It refers to the ability of systems to "see" when image processing algorithms identify entities depicted in images or videos with high speed and accuracy and then classify them. Similar applications can be found in robotics, security systems, navigation, surveillance systems, and a wide range of everyday human activities.[23]

Machine learning is an AI application that processes the generation of patterns or models from data sets. It can evolve and learn automatically to access databases for computer systems to learn automatically with autonomy and adjust their actions.[24] Using the right algorithms makes action planning and scheduling possible in several applications in industry and everyday life.

Artificial neural networks (Neural Networks, Parallel Distributed Processing Models, Connectionist Networks) mimic the groups of neurons in the human brain to process data and information quickly and flexibly.

A key feature of artificial neural networks is the ability to process information. Artificial neurons form large networks through the arrangement and connection in layers, where interconnections and learning ensure the functionality of the network system without being "programmed."[25]

The human brain inspired scientists to develop neural networks since the human brain has billions of neurons that communicate with each other, transmit electrical signals, and ensure the body's functioning through their connection to the body's cells. A fundamental characteristic of neural networks is the interconnectivity needed for information processing, which—despite the rapid advances in interconnectivity and processing power—is still inferior to the power of brain processing.[26]

The applications of artificial neural networks (ANN) are image recognition, scheduling, planning, voice processing, control, and production systems, and data mining with potential uses in industry, medical diagnostics, robotics, aviation, defense, automotive industry, geological research, economy and banking applications, and manufacturing in general.[27]

Moreover, AI exhibits great continuity and a wide range of development in Autonomous Robots, which can adapt and respond to external stimuli and have extended autonomy in different environments.[28]

Expert Systems can also contribute to decision-making or provide suggestions on specific issues.[29] Additionally, Knowledge Systems are software systems with intelligent adaptation to operational or problem-solving needs.[30]

A major and broad application of AI is intelligent agents, an intelligent system that processes information. This system can act empirically, combining information with perception. It can also effectively solve complex configurations based on data by selecting the best course of action to achieve the best outcome, resulting in several applications for everyday life.[31]

Solving problems requires the definition and design of an algorithm, which will walk through each problem until it is resolved through multiple courses of action and alternative algorithmic techniques throughout the development process and will determine the most appropriate option. When a problem is presented to an algorithm, an explicit description of the problem is given first, and as it evolves, it is defined so that it can serve as an input for the algorithm to process.[32]

One category of intelligent agents is Alexa and Siri, equipped with sensors to understand the request and then combine information gathered from global databases to reach a decision. The evolution of intelligent agents is rapid, as they have access to a larger volume of data, and at the same time, their functions are progressing with greater speed and precision.[33]

Examples of large-scale dialogue applications with an intelligent agent are Amazon's Alexa, Apple's Siri, and Microsoft's Cortana. As part of Windows 10 and macOS, they can interact in a natural language, which is also part of the NLP (Natural Language Processing) technology. Communication with these digital assistants has the characteristics and function of natural dialogues in response to questions and provision of solutions through data processing with rapid evolution and adaptation.[34]

Furthermore, "deep learning" is part of the machine learning framework that contributes to the understanding and optimization of system algorithms, computer algorithms that can learn and improve independently. It relates to neural network technology, which is used daily, for example, when browsing the Internet or using a mobile phone. As "deep learning" evolves, it becomes faster with greater accuracy, better efficiency, and limited supervision.[35]

Another fascinating area that has several benefits is the field of mindfulness and personal spiritual journeys. Here, a "guru" or a spiritual teacher can provide personalized guidance by analyzing your data (beliefs, background, walk of life). Moreover, that experience can be accelerated by connecting you with like-minded people in your area or region by creating a community of people sharing common experiences.

AI can analyze vast amounts of religious texts and scriptures, offering interpretations according to your current understanding and therefore assist individuals in deepening their faith in a spiritual path. Practically, it can generate a form of a "class for adults" who want to understand spirituality, how to improve themselves through mindfulness, and how to perceive the unseen world around them.

For this reason, AI can be used as a repository of religious knowledge, helping individuals navigate and access a vast storage of religious texts and scholarly works from diverse traditions. A move that democratizes access beyond geographical or language boundaries, providing information that was previously limited in its access.

Supposedly, a person uses advanced search tools. In that case, that user can leverage the power of sophisticated algorithms and identify patterns or receive answers that might be difficult for a person to detect. Humanity is winning here as it can analyze previously inaccessible data and receive answers directly for issues that transcend time and world localities.

Access to information was always seductive. Nowadays, facilitation generated by artificial intelligence makes researchers passively assimilate knowledge without actively employing analytical reasoning. The frenetic pace at which a human can process complex information, spending only a few seconds on each piece, results in negative consequences when dealing with deep theological subjects. Religious texts possess historical, cultural, and linguistic elements; without any doubt, meticulous examination and contemplation are necessary to understand them fully.

In addition to the above, AI translations are sometimes imperfect, particularly when dealing with complex religious issues. A fine example is how precisely a scholar can translate the term "theosis" from the Bible into English. It can have multiple interpretations and can easily be divinization, exaltation, or the process of becoming a God. In such a manner, it is very common to receive mistaken answers, so engineers must be prompted to examine the results and double-check if the outcome is correct. These algorithms, although remarkable, can make mistakes as humans do but also can encounter difficulties in capturing the cultural intricacies that exist in religious scripts.

Although AI is a powerful instrument for obtaining new knowledge, it should not be the sole compass for navigating religious texts. Such a profound understanding needs a thorough investigation and a discussion involving multiple interpretations.

## NOTES

1. Georgios Babiniotis, *Dictionary of Modern Greek* (Athens: Lexicology Center Ltd, 1998), p. 1196.
2. Stuart Russell, *Compatible: Artificial Intelligence and the Problem of Control* (New York: Penguin Random House, 2019), p. 3.
3. Pamela McCorduck, *Machines Who Think: A Personal Inquiry into the History and Prospects of Artificial Intelligence*, 2nd ed. (Natick, MA: A K Peters/CRC Press, 2004), pp. 10–25.
4. IEEE Global Initiative on Ethics of Autonomous and Intelligent Systems, "Ethically Aligned Design: A Vision for Prioritizing Human Well-Being with Autonomous and Intelligent Systems," First Edition (2019), https://ethicsinaction.ieee.org (08/01/2021).
5. UK House of Lords, Select Committee on Artificial Intelligence, "AI in the UK: Ready, Willing and Able?" *Report of Session* 2017–2019, (2018), [https://publications.parliament.uk/pa/ld201719/ldselect/ldai/100/100.pdf (10/12/2020)]. Mission assigned by the French Prime Minister, "For a Meaningful Artificial Intelligence: Toward a French and European Strategy," (2018). https://www.aiforhumanity.fr/pdfs/MissionVillani_Report_ENG-VF.pdf (10/12/2020).
6. Communication from the Commission to the European Parliament, the European Council, the Council, the European Economic and Social Committee and the Committee of the Regions on Artificial Intelligence for Europe, Brussels, 25.4.2018, COM (2018), 237, final. https://ec.europa.eu/futurium/en/system/files/ged/ai_hleg_definition_of_ai_18_december_1.pdf(12/12/2020).
7. Afraz Jaffri, with contributor Jackie Wiles, "What Is New in Artificial Intelligence from the 2022 Gartner Hype Cycle," Gartner, September 2022, https://www.gartner

.co.uk/en/articles/what-is-new-in-artificial-intelligence-from-the-2022-gartner-hype-cycle (15/09/2022).

8. Unknown Author, "Theories of Communication," University of Oklahoma, https://www.ou.edu/deptcomm/dodjcc/groups/99A2/theories.htm (15/09/2022).

9. Unknown Author, "Pygmalion," Encyclopaedia Britannica, https://www.britannica.com/topic/Pygmalion (15/09/2022).

10. A. M. Turing, "Computing Machinery and Intelligence," University of Oxford, 1950, pp. 433–460, https://www.cs.ox.ac.uk/activities/ieg/e-library/sources/t_article.pdf (21/09/2022).

11. European Commission's High-Level Expert Group on Artificial Intelligence, "Ethics Guidelines for Trustworthy AI," 143 (2019), https://ec.europa.eu/futurium/en/node/6945#_Toc414754698 (12/12/2020).

12. Nico Joost Kok, *Artificial Intelligence* (Oxford, UK: Eolss/UNESCO, 2009), p. 2. Stuart Russell and Peter Norvig, "Artificial Intelligence: A Modern Approach," trans. Takis Alvas, Dimitris Kartsaklis, and Fotis Skoularikis, ed. Klidarithmos, Athens 2005, p. 32. Ioannis Vlahavas, Petros Kefalas, Nikolaos Vasileiadis, Fotis Kokkoras, Ilias Sakellariou, *Artificial Intelligence* (Thessaloniki: University of Macedonia Press, 2006), p. 3.

13. M. Graves, "Embodied Experience in Socially Participatory Artificial Intelligence," *Zygon*, Vol. Unknown, 2023, pp. 3–4, https://doi.org/10.1111/zygo.12910 (15/09/2022).

14. Stuart J. Russell and Peter Norvig, "Artificial Intelligence: A Modern Approach," Pearson Education Limited, Vol. Unknown, 2021, pp. 1029–1033.

15. Themistocles Panagiotopoulos and Georgios Anastasakis, "Logic Programming," publ. University of Piraeus, 2012, pp. 2, 3, https://gunet2.cs.unipi.gr/modules/document/file.php/TME137/%CE%A3%CE%B7%CE%BC%CE%B5%CE%B9%CF% 8E%CF%83%CE%B5%CE%B9%CF%82/notes-lp-msc-inf-2012.pdf (15/12/2020).

16. European Commission's High-Level Expert Group on Artificial Intelligence, "Ethics Guidelines for Trustworthy AI," (2019) p. 144, https://ec.europa.eu/futurium/en/node/6945#_Toc414754708, (11/12/2020).

17. Georgios Krasadakis, "Artificial Intelligence: What It Is and How It Is Drastically Changing Our World," 3rd Naftemporiki Technology Conference, *Artificial Intelligence: A Blessing or a Threat*, Royal Olympic Hotel, 28 January 2018, Athens. https://www.naftemporiki.gr/story/1309187/texniti-noimosuni-ti-einai-kai-posallazei-dramatika-ton-kosmo-mas (10/01/2021).

18. Krasadakis, "Artificial Intelligence: What It Is and How It Is Drastically Changing Our World."

19. Vlahavas, Kefalas, Vasileiadis, Kokkoras, and Sakellariou, *Artificial Intelligence*, p. 4.

20. Katerina Georgouli, *Artificial Intelligence, an Introductory Approach*, Hellenic Academic Ebooks, Zografou 2015, pp. 21, 22. http://repfiles.kallipos.gr/html_books/93/01amain.html#_idTextAnchor002,file:///C:/User/User/Downloads/%CE%A4%CE%B5%CF%87%CE%BD%CE%B7%CF%84%CE%AE%20%CE%9DCE%BF

%CE%B7%CE%BC%CE%BF%CF%83%CF%8D%CE%BD%CE%B7%20(4).pdf (14/12/2020).

21. Georgouli, *Artificial Intelligence, an Introductory Approach*, pp. 21, 22.

22. Daniel Jurafsky and James Martin, *Speech and Language Processing: An Introduction to Natural Language Processing, Computational Linguistics and Speech Recognition*, 2020, pp. 487–489. https://web.stanford.edu/:jurafsky/slp3/ed3book.pdf (17/12/2020).

23. Krasadakis, "Artificial Intelligence: What It Is and How It Is Drastically Changing Our World"; Erik Miller, "Introduction to Computer Vision," University of Massachusetts Amherst, 2011, pp. 6, 7, https://people.cs.umass.edu/:elm/Teaching/Docs/IntroCV_1_19_11.pdf (19/12/2020).

24. Vlahavas, Kefalas, Vasileiadis, Kokkoras, and Sakellariou, *Artificial Intelligence*, pp. 15, 17.

25. Georgouli, *Artificial Intelligence, an Introductory Approach*, p. 155. Kenji Suzuki, *Artificial Neural Networks—Architectures and Applications* (Croatia: Intechopen, 2013), pp. 10–25.

26. Frederick Mills and Robert Stufflebeam, *Introduction to Intelligent Agents*, 2005 [https://miatedjosaputro.com/uploads/sites/2/2020/05/Introduction-to-Intelligent-Agents.pdf (12/12/20200]. Georgouli, *Artificial Intelligence, an Introductory Approach*, p. 155.

27. Georgouli, *Artificial Intelligence, an Introductory Approach*, p. 177.

28. Tom Harrys, "How Robots Work," howstuffworks.com, https://science.howstuffworks.com/robot4.htm (12/12/2020).

29. Georgouli, *Artificial Intelligence, an Introductory Approach*, pp. 200–205.

30. Ioannis Vlahavas, *Artificial Intelligence, Knowledge Systems*, 2013, opencourses .auth.gr, https://opencourses.auth.gr/modules/units/?course=OCRS118&id=1397 (12/12/2020).

31. Mills and Stufflebeam, *Introduction to Intelligent Agents*.

32. Georgouli, *Artificial Intelligence, an Introductory Approach*, p. 24.

33. Beth Hendricks, *Intelligent Agents: Definition, Types & Examples*, https://study.com/academy/lesson/intelligent-agents-definition-types-examples.html (11/12/2020).

34. Michael Bizzaco, Erika Rawes, and Kim Wetzel, *What Is Amazon's Alexa, and What Can It Do?* March 30, 2021. https://www.digital-trends.com/home/what-is-amazons-alexa-and-what-can-it-do/(10/04/2021). Chris Hoffman, *15 Things You Can Do with Cortana on Windows 10*, 2015. https://www.howtogeek.com/225458/15-things-you-can-do-with-cortana-on-windows-10/ (10/12/2020). Krasadakis, "Artificial Intelligence: What It Is and How It Is Drastically Changing Our World."

35. George Ceif, *Deep Learning vs Classical Machine Learning*, 2018, https://towardsdatascience.com/deep-learning-vs-classical-machine-learning-9a42c6d48aa (10/12/2020).

## *Chapter 2*

# Human-Centered Artificial Intelligence

Artificial intelligence (AI) is the study and design of algorithms that use data to perform tasks or exhibit behaviors that, if executed by a human, would require human intelligence. Also, intelligent systems are agents that can make some decisions on their own based on specific data and goals.[1]

Machine learning (ML) is an approach to intelligent system design in which the system adapts its behavior based on data. People increasingly interact with AI and ML systems, such as Siri, Alexa, Cortana, and Google Assistant. This is evident with self-driving cars but can also be less obvious, as with the "bank loan approval" algorithms that work behind the scenes.

Given the inherent potential of intelligent systems to affect people's lives, it is important to design intelligent systems with humanity at their epicenter. There is a growing awareness that algorithms are used in AI applications and that intelligent systems interact with humans. Human-centered AI gives the perspective that intelligent systems must be designed with the understanding that they are part of a larger system consisting of human components, such as users, operators, customers, and other individuals. Some researchers and AI professionals have been using the term "human-centered artificial intelligence" to describe intelligent systems developed with social responsibility in mind, addressing issues of justice, accountability, interpretation, and transparency. Human-centered AI can encompass more of these issues. Within this perspective, the broader scope of what it means to have a human-centered AI is explored, including factors governing our need for justice, interpretation, and transparency.[2]

Intelligent systems can solve problems, but they are incomprehensible to a person without prior computer science education. While humans are used to interacting with other humans and have developed abilities to predict their behavior, it is often not feasible to interact with intelligent systems, which do not solve problems as humans do. This is further exacerbated in the case of intelligent systems, where the user has no way of knowing which algorithms are being used or when the system is so complex that it is difficult to control.[3]

One way to avoid missteps toward the designated objectives is for logical systems to have common knowledge usually shared by individuals within the same society and culture. Common sense can also be useful in this instance, as can be observed, for example, in the case of driving cars on the right or left side of the road. Despite its shortcomings, efforts are being made to establish cognitive foundations of declarative rational knowledge based on different sources.[4]

Commonly accepted rational knowledge comes from sources created by humans, such as news, stories, dictionaries, encyclopedias, and, in general, open sources, along with the parallel definition of acceptable or unacceptable behaviors.[5]

Intelligent systems can use procedural knowledge to better provide services to individuals by predicting their behavior or detecting any misbehavior. In the same way, text prediction can also be helpful by anticipating larger patterns in everyday activities. Combining joint procedural knowledge with behavior can yield intelligent and safer AI functions. Rational procedural knowledge can assist intelligent systems in following social conventions to avoid conflict[6] and socially undesirable behaviors.

Human-centered artificial intelligence acknowledges that, for systems to be effective, they need the human mind's perspective. Just as human logic is used to interpret and predict the actions of others, intelligent systems can, in the same way, benefit from understanding the human reasoning. At the same time, it should also be pointed out that while sociocultural beliefs consciously or subconsciously influence people, intelligent systems are not "raised" in a society and culture the same way humans are.[7] For this reason, their software must be "taught" and adjusted accordingly.

Since sociocultural beliefs and rules provide a background that defines humans, any intelligent system should also be able to adapt to sociocultural beliefs and rules and decipher human behavior to anticipate and accommodate human needs. Thus, an intelligent system can better comprehend the sociocultural beliefs in human behavior, lowering the chances of making mistakes and becoming safer and more appealing to humans with increased compliance with moral rules.[8]

Some arguments can be found in the bibliography on the issue that AI systems with an anthropocentric orientation should be able to comprehend human sociocultural rules in conjunction with human cognition and provide clarification when an inexperienced end-user cannot understand and needs to be informed. In terms of ensuring the success of human-centered AI, further planning is required so that systems can better understand the human being.[9]

Given these facts, human-centered AI can be divided into two key modules: (a) the understanding of humans and (b) the ability of humans to understand AI systems. These two features could be the starting point of many desirable properties in intelligent systems interacting with inexperienced users and systems designed with social responsibility in mind. For instance, there is a growing awareness regarding the need for justice in advanced AI systems. Justice is the expectation that all users are treated equally and without bias. Therefore, a conscious effort is being made to gather data and develop system monitoring processes to prevent any harmful behavior.[10]

An intelligent system based on a population model of social and cultural norms with which it interacts can achieve justice and avoid discrimination and prejudice. At the same time, transparency is about providing certain means of access to data sets and work flows within a developing AI system with end-users as the recipients. The capacity to help people understand their decisions through explanations or other accessible means gives people greater trust and a better incentive to use AI systems. A human-centered AI does not necessarily mean an AI algorithm must think like a human being. On the other hand, as people are not AI experts, computer systems are designed to facilitate their interaction with humans based on sociocultural rules to avoid conflict between humans. Developing human-centered intelligent systems entails building intelligent systems based on common cultural criteria to meet people's expectations and needs. The development of human-centered AI focuses on the research for a scientific comprehension of the fundamental characteristics of AI while supporting the development of smart products and services that interact with civilians in real-life environments.[11]

António Guterres, secretary-general of the United Nations, stated in 2017: "As AI evolves into a benevolent tool for humankind, with the improvement of quality of life as its main guideline, this may result in a more sophisticated human species with a greater sense of integrity regarding governance, politics, and science, and integral compassion and empathy. This could bring all famines and human deaths, animal abuse, and the Earth's environmental destruction to an end. This could be the beginning of a planet that would be at peace. Artificial intelligence has the potential to accelerate progress towards a dignified life, in peace and prosperity, for all people. The time has arrived for all of us—governments, industry, and civil society—to consider how AI will affect our future."[12]

In 1637, in part V of his book *Discourse on the Method*, René Descartes considered whether an AI with full mental capacity could be created in the first place. He specifically states: "While intelligence is a universal instrument that can serve for all contingencies, [machines] have need of some special adaptation for every particular action. From this, it follows that it is impossible that there should be sufficient diversity in any machine to allow it to act in all the events of life in the same way as our intelligence causes us to act."[13]

In its procedural aspect, the commonly accepted logical knowledge can serve as a basis for the interaction with humans, to adjust human-AI interaction and render it more natural. Even if decision-making algorithms function in a different way than the human decision-making process, the system's behavior can and should be more adaptable to humans, providing safer interaction with humans and decreased objective failures, if the agent acts following humans' expectations and avoids conflicts by applying the theory of mind to human behavior.[14]

Besides the mildest applications, the challenges mentioned above can be overcome. Intelligence advances mechanical development through the optimal adaptation to social rules, customs, and values, so algorithmic applications can relate more easily to humans and limit the negative physical or psychological impacts that could cause social turmoil.[15]

In addition, there is a tendency to abandon the long-held notion that computers should act as human teammates, as they can allow designers to benefit from the discrete capabilities of algoıithms, databases, and sensors freely and more easily. For example, for managing satellites and other applications, such as the Mars Rovers, an entire room of human operators at NASA's Jet Propulsion Laboratory is dedicated to operating the Rovers as they perform planned tasks on Mars. Similarly, with surgical applications, such as the da Vinci Surgical Systems, a surgeon performs the surgery with the da Vinci system by using instruments guided through a console.[16]

Subsequently, according to Mumford[17], successful designs are not based on human forms. This challenges designers to understand which human features are useful, such as a human voice for virtual assistants like Siri, Alexa, and Cortana, which provide free access to information. However, designers realize that the visual display of long lists of text, a map, or diagrams is sometimes better than a verbal response. For example, the abundance of information in visual displays offers numerous advantages over a spoken response that might be less functional in a noisy environment. Apart from limiting design options, the humanization of robots can lead to misuse due to sentimental attachment to systems, false expectations of robotic responsibility, and misguided beliefs about appropriate robotic usage.[18]

Based on the above, researchers, trainers, designers, developers, managers and policymakers are creating new opportunities for developing future products and services, focusing on a high level of human control and automation. Therefore, the new approach to technological design differs from the conventional one-dimensional point of view that increased automation implies less human control[19] and contradicts the one-dimensional perspective of Sheridan and Verplank, although Sheridan[20] did express some ideas regarding the need for supervisory control.

Through the experience of applications, strategies for human-computer interaction[21] emerged, structured to balance automation and human control. In recent years, there has been a growing tendency towards more human control, even at high levels of automation, ensuring human supervisory intervention as automation also increases.[22]

The increasing interest in and use of artificial intelligence research has prompted extensive public and governmental scrutiny of the principles of ethics and responsibility to guide planners, managers, and policymakers. These principles are a helpful starting point, but bridging the gap between ethics and practice requires realistic management procedures, putting humans at the core of systems design thinking.[23]

In this two-dimensional context, human-centered artificial intelligence (HCAI)[24] demonstrates the design of automated systems in which humans maintain control. Examples of high levels of human control and automation include several well-known devices, such as thermostats or elevators, and life-critical applications, such as highly automated cars and patient-controlled pain relief devices. Smartphone cameras are an example of this new potential for human control and automation, where AI algorithms are used to adjust focus and aperture while regulating motion touch. Yet, users remain in control over the camera's placement, the best zoom level, and the timing of the shooting. They also control the filters before taking the photos and can edit, crop, and adjust the lighting afterward. Users can alter the aperture setting and exposure level by tapping on the part of the image they are interested in. However, a further useful user control would be to adjust the shutter speed to achieve longer exposures for artistic waterfall images or intentionally blurred photos of swirling dancers.[25]

What is also noteworthy is the tendency to shift from human emulation to the empowerment of humans. The two main objectives of AI development are imitating human behavior through AI science and developing useful applications through AI engineering. Thus, the "emulation" objective reinforced the notion that machines should be designed to resemble humans, whereas the "applications" objective can be best met by providing understandable, predictable, and auditable designs. While making computers intelligent, autonomous, and humane is now understood and accepted by researchers and designers, this effort must be further counterbalanced in terms of creating autonomous systems under supervisory control, considering the fact that many users want to be in charge of the technologies that support their abilities, contributing to efficacy, to the respect for their responsibility and the stimulation of their creativity.[26]

In addition, human-centered artificial intelligence involves bridging the gap between the socially accepted moral principles and the practical steps required for their implementation in governance structures by combining simulation-based designs. The trend is to adapt tried and tested techniques of group understanding and use of engineering software, as well as to apply management strategies throughout the design and manufacturing community to build a safety culture and establish independent supervision procedures that can be implemented to enhance performance.[27]

Management strategies for organizations begin with creating a safety culture driven by a leadership commitment that results in improved safety-oriented training and prevention practices. Other organizational strategies include detailed reports of failures and near misses, internal review boards for issues and plans, and alignment with standard application practices.

Based on software engineering practices and governance structures for human-centered AI, trustworthy systems are ensured through the control and availability of analysis tools and verification and confirmation trials. They are also secured by monitoring to strengthen justice, by leadership commitment to security, by security-oriented training, by extensive reporting of failures and near misses, by internal review boards for issues and plans, by ensuring government intervention and regulations, by involving non-governmental and civil society organizations, and by involving professional organizations and research institutions.[28]

These strategies lead to software development for HCAI products and services revolving around three goals: trustworthy systems relying on tried and tested software engineering practices with a group, security culture through organizational management strategies, and reliable certification by independent supervision. These recommendations are based on the human-centered AI framework by reducing the risks of excessive automation combined with human control while providing practice guidelines supporting the goal of safe and reliable products and services.[29]

Well-designed technologies can further support UN sustainable development goals, such as promoting high-quality education, gender equality, global peace, and social justice.[30]

## HUMAN-CENTERED ETHICS OF AI

AI is undertaking an ever-increasing range of interventions on behalf of humans in health, education, economy, security, transportation, governance, and infrastructure, as well as in many other aspects of daily personal activities. The extensive progressive use of AI raises many ethical dilemmas regarding its application in almost all public and private life areas.[31]

Despite the significant solutions provided by AI and the potential for progress in its application, several ethical dilemmas and questions with social and political implications arise. There is reasonable cause for concern regarding the impact on the labor market, as AI replaces many people's jobs and changes employment conditions, especially due to the collaboration of AI with robotics and its applications. Moreover, there are concerns regarding privacy protection and proper big data management as to how and who will process and use them, who will have access to the data, and who will analyze and be responsible for modeling them. The "training" and management of AI systems also raise questions.[32]

The ethical dilemmas emerge from the utilitarian application of AI, such as intelligent machines, robots, intelligent software, algorithms, smartphones, intelligent information systems, and the creation of intelligent artificial entities.

Let us illustrate this with an example: A big company receives tens of thousands of CVs yearly and must screen candidates and their skills before advancing them to the next stage. By using AI, they can implement parameters that the company works on—for example, teamwork, leadership skills, etc.—but they cannot understand the full extent of skills that an applicant has or the idiosyncrasies that arise from his cultural background. As a result, it might be possible that a good candidate will not pass the next stage as AI uses keywords and patterns that might not be present on their resume.

Another example of AI misuse is facial recognition, which has become present in many aspects of our social life. Nowadays, large public or private organizations have started using this technology to prevent fraud or to add another "weapon" in their cybersecurity armory. To give an example, a city council can utilize such a technology without clear ethical guidelines, and that may lead to misidentification, particularly towards minorities or people with an alien background. It is another subject that raises concerns about privacy and discrimination.

Here, the parameters utilized in governance must be crystal clear, as misuse and misinterpretations will tend to be the rule and not the exception. Additionally, foreign interference and disinformation can complicate things as these parameters need an "iron fence" protection from outsiders that want to disturb citizens' daily lives.

## NOTES

1. Mark Riedl, "Human-Centered Artificial Intelligence and Machine Learning," *Wiley School of Interactive Computing, Georgia Institute of Technology* (Atlanta, GA: 2018), p. 33 [DOI: 10.1002/hbe2.117 (18/12/2020)].

2. Riedl, "Human-Centered Artificial Intelligence and Machine Learning," p. 33.

3. Riedl, "Human-Centered Artificial Intelligence and Machine Learning," p. 34.

4. Douglas Lenat, "Cyc: A Large-Scale Investment in Knowledge Infrastructure," *Communications of the ACM* 38 (1995): 33–38. Ramakrishna Vedantam, Lin Xiaoyu, Batra Tanmay, L. C. Zitnick, and Devi Parikh, "Learning Common Sense through Visual Abstraction," paper presented at IEEE International Conference on Computer Vision (ICCV) (2015), 2542–2550. Fereshteh Sadeghi, Santosh Divvala, and Ali Farhadi. "VisKE: Visual Knowledge Extraction and Question Answering by Visual Verification of Relation Phrases," Paper presented at IEEE Conference on Computer Vision and Pattern Recognition (CVPR) (2015), 1456–1464.

5. H. Trinh Trieu and Trinh Quoc V. Le, "A Simple Method for Commonsense Reasoning," Cornell University (2018), pp. 2, 3, 8, 9 [CoRR; abs/1806.02847. https://arxiv.org/abs/1806.02847 (14/01/2021)].

6. Brent Harrison and Mark Riedl, "Towards Learning from Stories: An Approach for Interactive Machine Learning," Paper presented at the AAAI'16 Workshop on Symbiotic Cognitive Systems, 2016, 746–750 [https://www.cc.gatech.edu/:riedl/pubs/aaai-symbiotic16.pdf (14/01/2021)]. Riedl, "Human-Centered Artificial Intelligence and Machine Learning," p. 34.

7. Riedl, "Human-Centered Artificial Intelligence and Machine Learning," p. 34.

8. Tom Everitt, Victoria Krakovna, Laurent Orseau, and Shane Legg, "Reinforcement Learning with a Corrupted Reward Channel," Paper presented at the 2017 International Joint Conference on Artificial Intelligence (IJCAI), 2017, 4705–4713. Riedl, "Human-Centered Artificial Intelligence and Machine Learning," p. 34.

9. Riedl, "Human-Centered Artificial Intelligence and Machine Learning," p. 34.

10. Riedl, "Human-Centered Artificial Intelligence and Machine Learning," pp. 35, 36.

11. Riedl, "Human-Centered Artificial Intelligence and Machine Learning," pp. 35, 36.

12. United Nations, Secretary-General António Guterres, "UN Artificial Intelligence Summit Aims to Tackle Poverty, Humanity's 'Grand Challenges,'" 2017 [https://news.un.org/en/story/2017/06/ (02/01/ 2021)].

13. Rene Descartes, *Discourse on the Method Of Rightly Conducting the Reason, and Seeking Truth in the Sciences* [https://www.gutenberg.org/files/59/59-h/59-h.htm (10/01/2021].

14. Riedl, "Human-Centered Artificial Intelligence and Machine Learning," p. 34.

15. Mark Riedl, "Computational Narrative Intelligence: A Human-Centered Goal for Artificial Intelligence," *Georgia Institute of Technology*, Atlanta 2016 [https://arxiv.org/pdf/1602.06484.pdf (14/01/2021)]. Mark Riedl and Brent Harrison, "Using Stories to Teach Human Values to Artificial Agents. In Proceedings of the 2nd International Workshop on AI, Ethics and Society," *Georgia Institute of Technology* (Atlanta, GA: 2016) [https://www.cc.gatech.edu/:riedl/pubs/aaai-ethics16.pdf (17/01/2021)].

16. Davincisurgery, *Intuitive for Patients, Robotic-Assisted Surgery as a Minimally Invasive Option* [www.davincisurgery.com (14/01/ 2021)].

17. Lewis Mumford, *Technics and Civilization* (Chicago: University of Chicago Press 1934).

18. Lionel Robert, "The Growing Problem of Humanizing Robots," *International Robotics & Automation Journal* 3, no. 1 (2017): 247–248.

19. Robert, "The Growing Problem of Humanizing Robots," 247–248.

20. Thomas Sheridan, "Function Allocation: Algorithm, Alchemy or Apostasy?" *International Journal of Human-Computer Studies* 52, no. 2 (2000): 203–216.

21. Ben Shneiderman, *Designing the User Interface: Strategies for Effective Human-Computer Interaction* (Boston: Addison-Wesley, 1987), pp. 10–35.

22. Ben Shneiderman, "Human-Centered Artificial Intelligence: Reliable, Safe & Trustworthy," *International Journal of Human-Computer Interaction* 36, no. 6 (March 2020a): 495–504. Ben Shneiderman, Catherine Plaisant, Maxine Cohen, Steven Jacobs, Niklas Elmqvist, and Nicholas Diakopoulos, *Designing the User Interface: Strategies for Effective Human-Computer Interaction*, sixth ed. (Boston: Pearson, 2016), pp. 40–52.

23. Jessica Fjeld, Achten Nele, Hilligoss Hannah, Adam Nagy, and Madhulika Srikumar, "Principled Artificial Intelligence: Mapping Consensus in Ethical and Rights-Based Approaches to Principles for AI," *Berkman Klein Center Research Publication* (2020–2021) [https://dash.harvard.edu/bitstream/handle/1/42160420/HLS%20White%20Paper%20Final_v3.pdf (02/01/2021)]. Ben Shneiderman, "Human-Centered Artificial Intelligence: Three Fresh Ideas," *AIS Transactions on Human-Computer Interaction*, V.12 I.3 A.1, 9-30-2020, 120.

24. Ben Shneiderman, "Design Lessons from AI's Two Grand Goals: Human Emulation and Useful Applications," *IEEE Transactions on Technology and Society* 1, no. 2 (June 2020b): 73–82.

25. Shneiderman, "Human-Centered Artificial Intelligence: Three Fresh Ideas," 115.

26. Shneiderman, "Human-Centered Artificial Intelligence: Three Fresh Ideas," 116. Ben Shneiderman, "Design Lessons from AI's Two Grand Goals: Human Emulation and Useful Applications," *IEEE Transactions on Technology and Society* 1, no. 2 (June 2020b): 73–82.

27. Ben Shneiderman, "Opinion: The Dangers of Faulty, Biased, or Malicious Algorithms Requires Independent Oversight," *Proceedings of the National Academy of Sciences* 113, no. 48 (2016): 13538–13540. Ben Shneiderman, "Bridging the Gap between Ethics and Practice: Guidelines for Reliable, Safe, and Trustworthy Human-Centered AI Systems," *ACM Transactions on Interactive Intelligent Systems*, October 2020c [https://dl.acm.org/doi/10.1145/3419764 (03/01/2021)].

28. Shneiderman, "Human-Centered Artificial Intelligence: Three Fresh Ideas," 119–120.

29. Shneiderman, "Bridging the Gap between Ethics and Practice."

30. Shneiderman, "Human-Centered Artificial Intelligence: Three Fresh Ideas," 120.

31. Anastasia-Sofia Alexiadou, "Artificial Intelligence and the COVID-19 Pandemic. Moral Implications," *Archives of Hellenic Medicine* 37, no. 5 (2020): 709.

32. Krasadakis, "Artificial Intelligence: What It Is and How It Is Drastically Changing Our World."

## *Chapter 3*

# The Value of the Code of Ethical Conduct

To ensure proper operation and use, it is essential first to establish a clear definition of what constitutes an ethical code of conduct. Additionally, in this endeavor to define ethical principles, scientists designing, programming, and controlling artificial intelligent entities set ethical standards via "ethical algorithms" to attain morality attribution, which will affect practical applications.[1]

According to James Moor, human morality can be adapted to AI applications. Starting with the development of systems whose actions limit immoral outcomes, with systems that are programmed as to which actions they will and will not perform, as well as with the development of systems featuring fully moral entities that will be modeled after humans having free will, conscience, awareness, and sense of morality as much as possible.[2]

The applications of AI in both our private and public lives often affect or significantly influence decisions or even, in some cases, play a catalytic or independent role. The contribution in decision-making is determined based on the programming or algorithm regarding right or wrong, fair, or unfair, in conjunction with big data collection and private and public data analytics.[3] Given the above, these ethical dilemmas arise concerning the use of AI, and the ensuing outcomes.

The integration of ethics into AI occurs on three distinct levels. The first level, ethics by design, involves embedding ethical values into the system through carefully crafted algorithms. This approach ensures that the artificial autonomous system adheres to established ethical boundaries and standards, enabling it to make decisions that minimize or altogether avoid deviant behaviors and their potential consequences. The second level is about ethics in design. This means that methods, rules, and values are applied in combination with implementation rules, to underpin these values during the analysis and evaluation of the consequences and impact of the application of AI in the ethical sphere, either collectively or individually. The third level pertains to ethics for design, where the rules, criteria, and behaviors related to the reliability of manufacturing applications throughout the process and during all stages of research, design, implementation, development, and management of AI systems.[4]

Collectively, the ethical standards for AI systems are determined by the awareness (sentience) of the perception ability of the so-called perceptual experience, i.e., the perception of feelings and the awareness of wisdom (sapience), characterized by self-knowledge, logic, and intelligence.[5]

As AI systems evolve, they acquire the capacity for autonomous taking of initiatives, decision-making, and actions, raising the question of responsibility and accountability. This question is posed even though AI systems and applications are artifacts humans designed to achieve predefined goals.[6] In such a case, the methods and theories must be applied with the proper algorithms and their continuous adaptation, incorporating ethical values in their development, operation, redesign, and evaluation of the product or service delivered.

Consequently, an ethical code of conduct is imperative since rapid progress directly impacts humans and their environment. In terms of algorithmic transparency and accountability, computer algorithms are widely used in our economy and society to make decisions and have far-reaching implications. In their usefulness lies their demand for education, access to credit, healthcare, and employment.[7] Moreover, the pervasive role of algorithms in daily life underscores the necessity to confront challenges related to their design and technical aspects. This includes addressing potential side effects and preventing ethical biases right from their development.

At the same time, some scientists believe that AI is on an inevitable path toward relative or absolute autonomy. For decades, artificial intelligence has been evolving rapidly. Nowadays, computer systems can read X-rays, make medical diagnoses, fly airplanes, and analyze forensic data. Algorithms can paint masterpieces and compose symphonies. Google is developing "artificial moral reasoning" to enable its self-driving cars to make decisions in the face of potential accidents.[8]

An algorithm is a self-contained, step-by-step set of operations that computers and other "smart" devices use to perform calculations, data processing, and automated reasoning tasks. Algorithms increasingly implement institutional decision-making based on analysis, which involves detecting, interpreting, and communicating meaningful patterns in data. Particularly valuable in data-rich environments with recorded information are analytics that rely on the simultaneous use of statistics, computation, programming, and operations research to quantify performance. There is also growing evidence that some algorithms and analytics may be non-transparent, making it impossible to determine when their outputs may be biased or incorrect.[9]

Computational models can be distorted due to biases embedded in their input data or algorithms. Decisions made by predictive algorithms may be non-transparent because of numerous factors, such as technical factors (the algorithm may not be amenable to easy explanation), financial (provision costs, transparency may be excessive, including the compromise of trade secrets), and social (the disclosure process may violate expectations of confidentiality). Even well-designed computer systems can lead to unexplained outcomes or errors, either because they contain bugs or because the circumstances of their use change, invalidating the premises on which the original analytics were based.[10]

Using algorithms in automated decision-making regarding individuals can lead to harmful effects. Policymakers should maintain institutions that use analytical data based on the same patterns as those of people who have traditionally made relevant decisions about their daily needs. Developers should design analytical systems to adhere to these standards when algorithms are used to make automated decisions or when algorithms offer input to human decisions. According to the ACM Code of Ethics, this set of principles is intended to advocate the merits of algorithmic decision-making when addressing people's and communities' everyday problems. These principles should be adopted during each phase of system design and development to the extent necessary to minimize potential moral injury, recognizing the benefits of algorithmic decision-making.[11]

The algorithmic transparency and accountability principles are Awareness, Access and redress, Accountability, Explanation, Data Provenance, Auditability, Validation, and Testing. Through the principle of Awareness, owners, designers, builders, users, and other stakeholders of analytic systems should be aware of the potential biases involved in their design, implementation, and use and the potential harm biases can cause to individuals and society. Through Access and redress, Regulatory Authorities should encourage the adoption of mechanisms that allow for restitution and compensation for individuals and groups adversely affected by algorithmically informed decisions.[12]

Accountability is also essential so that the institutions can be held responsible for the decisions made by the algorithms they use, even if it is not feasible to explain in detail how the algorithms produce their results. Subsequently, in the context of public policy, systems and institutions that use algorithmic decision-making must define the procedures followed by the algorithm for decision-making.[13]

The way data are collected and stored by algorithm developers can potentially harbor biases, stemming from either human involvement or the algorithmic data collection process itself. There are privacy concerns, such as safeguarding trade secrets or disclosing analytical data, that could allow malicious actors to gain access when no provision has been made to restrict access to qualified and authorized individuals. At this point, public scrutiny of the data provides the potential and the opportunity for corrections. Algorithms, models, data, and decisions must be documented so that they can be reviewed in cases where there is a suspicion of malfunction. Institutions should perform regular audits to assess and determine whether the algorithmic models produce discriminatory errors and employ rigorous methods to validate their algorithmic models to document the methods and outcomes.[14]

As artificial intelligence continues to develop, the issue of whether it can have moral agency becomes increasingly significant. In brief, moral agency is the capacity to make moral decisions, accept responsibility, and be subject to moral obligations. In artificial intelligence, moral agency raises issues regarding whether a computing system can possess ethical responsibility for its actions and how this responsibility should be assigned and evaluated.

One perspective on moral agency is anchored in "machine ethics," which aims to develop AI systems capable of autonomously making ethical decisions. Machine ethics posits that these systems, as they grow more sophisticated and autonomous, could possess a set of morals and values, given that they can make complex decisions with ethical implications. To ensure the behavior of artificial intelligence systems, proponents of machine ethics argue that we must develop methods to teach and incorporate values and principles into their programming.

It is necessary to examine moral reasoning to investigate that possibility. Moral reasoning encompasses the cognitive processes by which individuals form ethical judgments and decisions. There are numerous moral reasoning theories, such as deontological ethics, consequentialism, and virtue ethics. Each of these theories provides a unique viewpoint on how ethical decisions should be made, which can be applied to the context of artificial intelligence.

Deontological ethics, for instance, asserts that moral decisions should be based on rules or duties, regardless of their consequences. In the context of artificial intelligence, this would entail the creation of AI systems that adhere to a predetermined set of ethical rules or principles. In contrast, consequentialism asserts that the morality of an action is determined by its consequences. In this framework, AI systems must be able to evaluate and predict the outcomes of their moral decision-making efforts. Finally, virtue ethics emphasizes a person's moral character rather than norms or consequences, implying that AI systems require moral virtues to behave ethically.

Nowadays, as AI technologies infiltrate many aspects of our daily lives, the issue of ethical decision-making has become a major long-standing challenge. An AI actor can be considered ethical if it weighs the consequences of each choice and picks the most ethical outcomes or decisions that can be achieved while often aiming to deliver the best overall service.

An actor is ethical if he/she complies with the obligations, duties, and rights associated with specific situations and follows established social norms with moral values in human-AI interaction.[15]

Artificial intelligence and robotics face an important ethical dilemma, whether to support or oppose the development of lethal autonomous weapons, as the technology for such systems already exists. The issue is whether autonomous weapon systems will select and engage targets without human intervention and whether lethal strikes on targets will involve humans. Thus, they can seek and destroy enemy combatants on a battlefield or city with cruise missiles or drones, for which, up until now, humans have made all the decisions regarding targets. Existing AI and robotics elements can provide platforms, perception, engine control, navigation, mapping, tactical decision-making, and long-term planning. In such cases, the fundamental principles of human dignity can be violated by allowing machines to choose who they will kill if they are instructed to eliminate anyone with a "threatening behavior."[16]

For example, according to a published UN report, in 2020, an STM Kargu-2 unmanned military aircraft operating with artificial intelligence attacked a group of armed men in Libya without human control. The report describes it as a "lethal autonomous weapon system," operating on artificial intelligence and used by the forces supporting the Tripoli government against rival armed forces, trying to evade missile attacks. It is stated that the weapon system "was programmed to attack targets without requiring data connectivity between the operator and the ammunition."[17]

## CODES OF ETHICAL CONDUCT IN RELEVANCE WITH HUMAN VALUES AND HUMAN RIGHTS

The principles established in the Belmont Report of 1978 for the behavioral and technological sciences constitute a starting point for ensuring ethics. The principles encompass three basic requirements: (1) personal autonomy should not be violated, they should be able to maintain their free will when interacting with technology; (2) the technological benefits should outweigh the risks; and (3) benefits should be allocated fairly among users and without discrimination based on any personal background, such as race, gender, and religion.[18]

In April 2019, the commission referred the communication to the European Parliament, the Council, the European Economic and Social Committee, and the Committee of the Regions "Building Trust in Human Centric Artificial Intelligence." The European Union (EU) commissioned a high-level group of experts to draw up guidelines regarding the credibility of AI. The guidelines drafted by the expert group on AI, on the issue of achieving a "trustworthy

AI," set out three components: AI should comply with the law, respect ethical principles, and be competent. These three components are defined by nine key requirements, which AI applications should meet for them to be considered trustworthy: they should fulfill the criteria of human intervention and oversight, transparency, accountability, social and environmental well-being, technical excellence and safety, diversity, non-discrimination, fairness, and privacy and data protection.[19]

In the European Parliament Resolution of January 20, 2021, the texts regarding AI in questions of interpretation and application of international law, in so far as the Union is affected in the areas of military and non-military use and matters of state authority beyond the scope of criminal justice, were adopted.[20] In the section on public international law and on military uses of AI, the resolution states that the AI applications used in a military and non-military context must be subject to meaningful human control so that at any given time, a person has the power to stop, correct, or turn off the AI-based technology in the event of accidental intervention, unpredictable behavior, cyber-attacks, or third-party interference.[21]

Paragraph four underlines the respect for humanitarian, international, and public law, which applies to AI weapon systems and their operators, emphasizing that the Member States must comply, especially in situations where they are protecting the civilian population or taking preventive measures in the event of an attack, cyberwarfare, or military attack.[22]

In any autonomous decision-making by AI systems, humans should not be absolved of the ultimate responsibility in the decision-making process to ensure that the person responsible for the decision can be identified.[23]

During the use of AI by EU Member States in a military context, parties to a conflict and individuals must comply with international law, with a commensurate assumption of responsibility for actions resulting from the use of such systems; likewise, adverse actions, anticipated or accidental effects of the AI systems are the responsibility of the parties of a conflict of states or individuals.[24]

EU encourages designing, developing, testing, and implementing AI systems but stipulates that potential risks must be considered. It also advocates for training applications in both military and civilian AI systems, as well as for processing health data for monitoring and ensuring the well-being of personnel.[25]

The promotion of supporting certification and surveillance systems through explicit mechanisms of explainability, auditability, traceability, and accountability is also encouraged. All these elements are important and set in motion in a hyper-connected world, an international legal framework for artificial intelligence, with the EU taking the lead in this area and taking initiatives in coordination with the international community and the United Nations. This international legal framework must ensure the use of AI for military and other purposes within the limits set by international law, particularly the Geneva Conventions of August 12, 1949, and international humanitarian law.[26]

With the ability of AI to act as a force multiplier for any system in which it is used, promoting human values is a key element of ethics and respect for rights. The principles of promoting human values are in 69 percent of the documents. The principles correspond to and are influenced by social norms. As the use of AI becomes more widespread and the power of technology and artificial intelligence increases, implementing human priorities is particularly crucial. Promoting human values is guided by three principles: human values and well-being, access to technology, and its use for the benefit of society. These principles regarding human rights were established since human values and rights are very much in line with international human rights law, which guarantees human dignity and autonomy. The Organization for Economic Co-operation and Development (OECD) and the G20 have included the principles of artificial intelligence as “internationally recognized labor rights."[27]

Following the principle of utilizing artificial intelligence for the benefit of society, to serve people, and for them to be in an operational process, AI systems should remain compatible with human principles, namely dignity, integrity, freedom, protection of privacy and culture, and gender diversity, as well as fundamental human rights. Furthermore, AI systems must protect and contribute to the common effort to improve our planet’s ecosystems and biodiversity.[28]

The principle of human value and well-being states that the development and use of artificial intelligence should be about the prevailing social networking, rules, and fundamental cultural beliefs and work towards improving humanity. AI should “serve the advancement of human civilization".[29]

A special reference regarding human values is made in the Montreal Declaration so that AI systems can promote people’s well-being and improve their living conditions, health, employment, and general conditions that enable people to carry out their mental and physical activities. In contrast, AI should not contribute to increased anxiety or the impression that digital environments harass people.[30] Moreover, an equal distribution of wealth is necessary, so that the wealth created by AI will benefit both workers and society as a whole.[31] Regarding AI, the Japanese authorities acknowledge its economic aspect, pointing out that AI should not result in a setting where wealth and social influence are unduly biased against certain citizens.[32]

The socio-political dimensions of inequality are highlighted, including the potential for AI to unfairly benefit certain countries or regions due to the digital divide caused by “information-deprived” or “technology-deprived” people.[33]

Some views on "access to technology" are based on the notion that wide access to AI technology refers to the education associated with it, which is necessary for its proper use and understanding. The Chinese AI governance principles underline that AI stakeholders should be able to receive education and training to assist them in adjusting to the implications of AI development along with the corresponding psychological, emotional, and technical aspects.[34]

The EU high-level expert group guidelines reflect a vision of "accessibility" to this technology for people with disabilities, noting that in some cases, AI-enabled services empower individuals with hearing, visual, and other impairments.[35]

The Information Technology Industry Council's policy principles on AI focus on educating people who have traditionally been marginalized or excluded from technological innovation, calling for diversifying and expanding access to the resources needed for AI development and use.[36]

The principle of leverage towards the benefit of society assumes that AI is utilized in such a way that it benefits society in the sense that the systems serve the community's goals. These objectives include social, political, and economic factors. Some examples of beneficial purposes set out in the guidelines of the European high-level expert group are respect for human dignity, individual freedom, respect for democracy, justice and the rule of law, equality, non-discrimination, and solidarity—including the rights of persons at risk of exclusion, citizens' rights, including the right to vote, the right to good administration or access to public documents and the right to petition.[37] The high-level expert group and the German AI strategy were the two documents that explicitly include the environment and sustainable development as factors in the definition of AI that is "leveraged to benefit society."[38]

Not only must AI systems be lucrative, secure, and legal, but they must also include human well-being as a primary success criterion for development.[39] According to Google, the company will move forward with the development of AI technology to achieve an overall benefit that significantly exceeds the foreseen risks and drawbacks after considering the wide range of social and economic factors.[40]

Regarding human control over the technology, the European Commission's high-level expert group on artificial intelligence, led by the concerns of societies, governments, and corporations, addresses the potential of humans controlling AI.

The literature on human control of AI converges on the principles of "human revision of the automated decision, the option of opting out of the automated decision and the human control of the technology." Human revision can contribute towards the objectives of security, transparency and explainability, and justice and non-discrimination by promoting human values.[41] For instance, the OECD and G20 principles on AI refer to human scrutiny as "safeguarding,"[42] and the UNI Global Union claims that transparency in decisions and outcomes necessitates "the right to contest decisions made by AI/algorithms, and to human scrutiny."[43]

The principle of "human revision of the automated decision" points out that when AI systems are implemented, individuals subject to these systems' decisions should be able to request and receive human revision of these decisions. Unlike other principles, in the context of this issue, the "human revision of the automated decision" principal functions as an ex-post way of correcting an erroneous result. The reasoning is that "people interact with AI systems and must be able to maintain full and effective self-determination.[44]

The European Charter on artificial intelligence in judicial systems also includes a strong version of the principle of human revision, specifying that if a revision is sought, it must be valid and reviewed by a competent court. There is also the "option of opting out of the automated decision," offering individuals the opportunity and choice not to be subject to applied AI systems.[45]

In the UK, it is a requirement that the public be aware of how and when AI is being used in the decision-making process and what the implications are for each person individually. This clarity, combined with increased digital education, will help the public better understand the benefits of AI and the exemption of using such products in cases where they raise concerns.[46]

"Human control over technology," as a principle, required the design of AI systems and was implemented with the possibility for humans to intervene in their actions. The Asilomar issue of AI principles states that humans must decide how and whether they will delegate decisions to AI systems to achieve human-selected goals while promoting human values and quality of life.[47] The Montreal Declaration states that we must create AI systems that respect people's autonomy while increasing control over their lives and their environment[48] through legal instruments and entities so that control and responsibility are always maintained.[49]

The relationship between artificial intelligence and human rights is inherently intricate. A single AI application can simultaneously impact several political, economic, social, and cultural rights, yielding positive and negative outcomes for diverse groups. Like historical precedents, the global community is now striving to strike a balance amidst the myriad challenges posed by AI applications.

In our time, humanity is in an advantageous position compared to the past due to the adoption, in December 1948, of the Universal Declaration of Human Rights (UDHR) by the UN.[50]

The UDHR provides a solid and universally accepted framework for the definition of human rights, overcoming past mistakes and ensuring respect for the rights of every human being. The guiding principles help set the boundaries regarding using new technologies with the necessary adjustments.

There are human rights–based processes at a global level for assessing and addressing the social impacts of AI applications. In these processes, the role of governments and international organizations is crucial in providing the available remedies in cases where human rights are violated. Thus, assessing and addressing the consequences of AI is part of the governmental responsibility. Institutions, procedures, and democratic governance determine the legitimacy of defining the fair distribution of benefits throughout society.[51]

## NOTES

1. Nick Bostrom and Eliezer Yudkowsky, "The Ethics of Artificial Intelligence," in *The Cambridge Handbook of Artificial Intelligence*, edited by K. Frankish, W. M. Ramsey (Cambridge: Cambridge University Press, 2011), pp. 316–334 [https://www.fhi.ox.ac.uk/bostrom-n-the-ethics-of-artificial-intelligence%C2%94-in-the-cambridge-handbook-of-artificial-intelligence-eds-william-ramsey-and-keith-frankish-cambridge-cambridge-university-press-2014-w/ (11/11/2020)]. Virginia Dignum, "Ethics in Artificial Intelligence: Introduction to the Special Issue. Ethics and Information Technology," *Springer, Ethics and Information Technology* 20 (2018): 1–3 [https://link.springer.com/article/10.1007/s10676-018-9450-z (11/11/2020)]. Alexiadou, "Artificial Intelligence and the COVID-19 Pandemic. Moral Implications," p. 710.

2. James Moor, "The Nature, Importance, and Difficulty of Machine Ethics," IEEE Intelligent Systems, 2006, 18, 21 [doi: 10.1109/MIS.2006.80 (11/12/2020)].

3. Alexiadou, "Artificial Intelligence and the COVID-19 Pandemic. Moral Implications," pp. 709, 710.

4. Dignum, "Ethics in Artificial Intelligence: Introduction to the Special Issue. Ethics and Information Technology," pp. 1–3. Kaira Sekiguchi and Koichi Hori, "Organic and Dynamic Tool for Use with Knowledge Base of AI Ethics for Promoting Engineers' Practice of Ethical AI Design," *Springer, AI & Society, Journal of Knowledge, Culture and Communication* 35, no. 51–71 (2020) [https://link.springer.com/article/10.1007/s00146-018-0867-z, (10/11/2020)]. Alexiadou, "Artificial Intelligence and the COVID-19 Pandemic. Moral Implications," p. 710.

5. Bostrom and Yudkowsky, "The Ethics of Artificial Intelligence," pp. 318–330.

6. Dignum, "Ethics in Artificial Intelligence: Introduction to the Special Issue. Ethics and Information Technology," pp. 1–3. Mark Coeckelbergh, "Virtual Moral Agency, Virtual Moral Responsibility: On the Moral Significance of the Appearance, Perception, and Performance of Artificial Agents," *Springer, AI & Society, Journal of Knowledge, Culture and Communication* 24 (2009): 181–189 [https://link.springer.com/article/10.1007%2Fs00146-009-0208-3 (11/11/2020)].

7. Federal Trade Commission, "Big Data: A Tool for Inclusion or Exclusion? Understanding the Issues," January 2016, pp. 1, 2 [https://www.ftc.gov/system/files/documents/reports/big-data-tool-inclusion-or-exclusion-understanding-issues/160106big-data-rpt.pdf (14/11/ 2020)].

8. Aimee van Wynsberghe and Scott Robbins, "Critiquing the Reasons for Making Artifcial Moral Agents," *Springer, Science and Engineering Ethics* (June 2019): 719–735.

9. Bostrom and Yudkowsky, "The Ethics of Artificial Intelligence," pp. 320–334. Mark Coeckelbergh, "Virtual Moral Agency, Virtual Moral Responsibility: On the Moral Significance of the Appearance, Perception, and Performance of Artificial Agents," *Springer* 24 (2009): 181–189 [https://link.springer.com/article/10.1007/s00146-009-0208-3(20/02/2021)].USPublicPolicyCouncil (USACM), Association for Computing Machinery, "Statement on Algorithmic Transparency and Accountability," p. 1 [https://www.acm.org/binaries/content/assets/ public-policy/2017_usacm_statement_algorithms.pdf (15/11/2020)].

10. US Public Policy Council (USACM), Association for Computing Machinery, "Statement on Algorithmic Transparency and Accountability," p. 1.

11. US Public Policy Council (USACM), Association for Computing Machinery, "Statement on Algorithmic Transparency and Accountability," p. 1.

12. US Public Policy Council (USACM), Association for Computing Machinery, "Statement on Algorithmic Transparency and Accountability," p. 2.

13. US Public Policy Council (USACM), Association for Computing Machinery, "Statement on Algorithmic Transparency and Accountability," p. 2.

14. US Public Policy Council (USACM), Association for Computing Machinery, "Statement on Algorithmic Transparency and Accountability," p. 2.

15. Wendell Wallach and Allen Colin, *Moral Machines: Teaching Robots Right from Wrong* (Oxford: Oxford University Press, 2009), pp. 2–6.

16. Russell Stuart, "Take a Stand on AI Weapons," *Nature* 521 (May 2015): 415–418 [https://www.nature.com/articles/521415a. pdf?origin=ppub (15/11/2020)].

17. Maria Cramer, "A.I. Drone May Have Acted on Its Own in Attacking Fighters, U.N. Says," *New York Times*, June 3, 2021 [https://www.nytimes.com/2021/06/03/world/africa/libya-drone.html (17/07/2021)]. David Hambling, "Drones May Have Attacked Humans Fully Autonomously for the First Time," *newscientist*, May 27, 2021 [https://www.newscientist.com/article/2278852-drones-may-have-attacked-humans-fully-autonomously-for-the-first-time/ (17/07/2021)].

18. Rose Luckin, "Towards Artificial Intelligence Based Assessment Systems," *Nature Human Behaviour*, volume 1, art, n. 0028, 2017, pp. 1–3 [https://www.nature.com/articles/s41562-016-0028?proof=tNature—doi:10.1038/s41562–016–0028 (18/12/2020)]. Han Yu, Shen Zhiqi, Miao Chunyan, Cyril Leung, Victor Lesser, and Qiang Yang, *Building Ethics into Artificial Intelligence*, 2018 [https://arxiv.org/abs/1812.02953 (17/12/2020)].

19. European Commission, Communication from the Commission to the European Parliament, the Council, the European Economic and Social Committee and the Committee of the Regions, *Building Trust in Human-Centric Artificial Intelligence*,

8.4.2019 Brussels, pp. 4–8 [https://ec.europa.eu/transparency/documents-register/detail?ref=COM(2019)168&lang=el (15/12/2020)].

20. European Parliament 2019–2024, "Texts Adopted, Artificial Intelligence: Questions of Interpretation and Application of International Law, European Parliament Resolution of 20 January 2021 on Artificial Intelligence: Questions of Interpretation and Application of International Law in So Far as the EU Is Affected in the Areas of Civil and Military Uses and of State Authority Outside the Scope of Criminal Justice," 2020, p. 1 [https://www. europarl.europa.eu/doceo/document/TA-9-2021-0009_EL.pdf (09/01/2021)].

21. European Parliament 2019–2024, "Texts Adopted, Artificial Intelligence: Questions of Interpretation . . . Criminal Justice" § 3.

22. European Parliament 2019–2024, "Texts Adopted, Artificial Intelligence: Questions of Interpretation . . . Criminal Justice" § 4.

23. European Parliament 2019–2024, "Texts Adopted, Artificial Intelligence: Questions of Interpretation . . . Criminal Justice" § 8.

24. European Parliament 2019–2024, "Texts Adopted, Artificial Intelligence: Questions of Interpretation . . . Criminal Justice" § 9.

25. European Parliament 2019–2024, "Texts Adopted, Artificial Intelligence: Questions of Interpretation . . . Criminal Justice" § 10–13.

26. European Parliament 2019–2024, "Texts Adopted, Artificial Intelligence: Questions of Interpretation . . . Criminal Justice" § 14–15.

27. Jessica Fjeld, Nele Achten, Hannah Hilligoss, Adam Nagy, Madhulika Srikumar, "Principled Artificial Intelligence: Mapping Consensus in Ethical and Rights-Based Approaches to Principles for AI," p. 60. Organisation for Economic Co-operation and Development, "Recommendation of the Council on Artificial Intelligence," n. 54, Principle 1.2, p. 7 [https://legalinstruments.oecd.org/en/instruments/OECD-LEGAL-0449 (03/03/2021)]. G20 Trade Ministers and Digital Economy Ministers, "G20 Ministerial Statement on Trade and Digital Economy," 2019, Principle 1.2, p. 11 [https://www.mofa.go.jp/files/ 000486596.pdf (03/03/2021)].

28. Future of Life Institute, "Asilomar AI Principles," 2017, n. 89 [https://futureoflife.org/ai-principles/?cn-reloaded=1 (03/02/2021)]. Council of Europe, European Commission for the Efficiency of Justice, "European Ethical Charter on the Use of Artificial Intelligence in Judicial Systems and Their Environment," 2018, Principle 1, p. 8 [https://rm.coe.int/ethical-charter-en-for-publication-4-december-2018/16808f699c (03/02/2021)]. UNI Global Union, "Top 10 Principles for Ethical Artificial Intelligence" 2017, n. 66, Principle 3, p. 7 [http://www.thefutureworldofwork.org/media/35420/uni_ethical_ai.pdf (03/02/2021)].

29. Chinese National Governance Committee for the New Generation Artificial Intelligence, led by China's Ministry of Science and Technology, "Governance Principles for a New Generation of Artificial Intelligence: Develop Responsible Artificial Intelligence," 2019, n. 22, Principle 1 [http://www.chinadaily.com.cn/a/201906/17/ WS5d07486ba3103dbf14328ab7.html (04/02/2021)].

30. University of Montreal, "2018 Report Montréal Declaration for a Responsible Development of Artificial Intelligence," 2018, n. 34,

Principle 1, p. 8 [https://monoskop.org/images/b/b2/Report_Montreal_Declaration_for_a_Responsible_Development_of_Artificial_Intelligence_2018.pdf(06/02/2021)].

31. Tencent Institute, "Six Principles of AI," n 58, Principle 1, [http://www.kejilie.com/iyiou/article/ZRZFn2.html (06/02/2021)].

32. Japanese Cabinet Office, Council for Science, Technology and Innovation, "Social Principles of Human-Centric Artificial Intelligence," n. 20, 2019, p. 9 [https://www8.cao.go.jp/cstp/english/human-centricai.pdf (06/02/2021)]. Microsoft, "AI Principles," 2018 [https://www.microsoft.com/en-us/ai/our-approach-toai> (09/02/ 2021)].

33. Japanese Cabinet Office, Council for Science, Technology and Innovation, "Social Principles of Human-Centric Artificial Intelligence," n. 20, 2019, p. 7.

34. Chinese National Governance Committee for the New Generation Artificial Intelligence, led by China's Ministry of Science and Technology, "Governance Principles for a New Generation of Artificial Intelligence: Develop Responsible Artificial Intelligence," n. 22, Principle 2.3.

35. European Commission's High-Level Expert Group on Artificial Intelligence, "Ethics Guidelines for Trustworthy AI" (2019), pp. 18, 36.

36. Information Technology Industry Council, "AI Policy Principles" (2017), n. 9, p. 5 (05/12/2020).

37. European Commission's High-Level Expert Group on Artificial Intelligence, "Ethics Guidelines for Trustworthy AI" (2019), n. 6, p. 11, Principle 2.1, p. 11.

38. European Commission's High-Level Expert Group on Artificial Intelligence, "Ethics Guidelines for Trustworthy AI" (2019), n. 6, Principle 2.1, p. 11. n. 5 p. 32. German Federal Ministry of Education and Research, the Federal Ministry for Economic Affairs and Energy, and the Federal Ministry of Labour and Social Affairs, "Artificial Intelligence Strategy." (2018) n 9, p. 9 [https://www.ki-strategie-deutschland.de/home.html (06/12/2020)].

39. IEEE Global Initiative on Ethics of Autonomous and Intelligent Systems, "Ethically Aligned Design: A Vision for Prioritizing Human Well-Being with Autonomous and Intelligent Systems," n. 5, Principle 2, pp. 21–22.

40. Fjeld, Acheten, Hilligoss, Nagy, and Srikumar, "Principled Artificial Intelligence: Mapping Consensus in Ethical and Rights-based Approaches to Principles for AI," pp. 24, 56.

41. Fjeld, Achten, Hilligoss, Nagy, and Srikumar, "Principled Artificial Intelligence: Mapping Consensus in Ethical and Rights-based Approaches to Principles for AI," p. 52.

42. University of Montreal, "2018 Report Montréal Declaration for a Responsible Development of Artificial Intelligence," 2018, n. 34, Principle 2, p. 9.

43. Japanese Cabinet Office, Council for Science, Technology and Innovation, "Social Principles of Human-Centric Artificial Intelligence," 2019, n 20, p. 17.

44. University of Montreal, "2018 Report Montréal Declaration for a Responsible Development of Artificial Intelligence," 2018, n. 34, Principle 9, p. 9.

45. European Commission's High-Level Expert Group on Artificial Intelligence, "Ethics Guidelines for Trustworthy AI" (2019), n. 6, Principle 2.1, p. 11.

46. UK House of Lords, Select Committee on Artificial Intelligence, "AI in the UK: Ready, Willing and Able?" (2018) *Report of Session* 2017–19 (n. 8), p. 27.

47. Japanese Cabinet Office, Council for Science, Technology and Innovation, "Social Principles of Human-Centric Artificial Intelligence," 2019, n. 20, p. 4. Future of Life Institute, "Asilomar AI Principles," 2017, n. 89, Principle 16.

48. University of Montreal, "2018 Report Montréal Declaration for a Responsible Development of Artificial Intelligence," 2018, n. 34, Principle 2, p. 9.

49. UNI Global Union, "Top 10 Principles for Ethical Artificial Intelligence," 2017, n. 65, Principle 4, p. 8.

50. United Nations, "Universal Declaration of Human Rights," December 10, 1948 [https://www.un.org/en/about-us/universal-declaration-of-human-rights (03/12/2020)].

51. Filippo Raso, Hannah Hilligoss, Vivek Krishnamurthy, Christopher Bavitz, and Kim Levin, "Artificial Intelligence & Human Rights: Opportunities and Risks," *Berkman Klein Center Research Publication*, September 25, 2018, p. 58 [https://cyber.harvard.edu/ sites/default/files/2018-09/2018-09_AIHumanRightsSmall.pdf (03/12/2020)].

# *Chapter 4*

# AI and Governance

The literature on the ideal AI governance[1] explores the potential international arrangements for managing the concerns and worldviews of people from different backgrounds. Governance proposals are more likely to have a global appeal and be compatible with stakeholder incentives. In governance, the more AI is part of it, the more a realistic institutional design, constitutionally based, is needed to define who is entitled to make which decisions and under which conditions. It should also stipulate and determine what information is communicated to whom to safeguard democratic processes.

The ideal AI governance aspires to design and promote optimal institutional solutions to the challenges of humanity's governance on shared values and regarding which various groups the AI resources can coordinate. There are many different groups of interested parties, such as civil society, artificial intelligence researchers, governments, corporations, and other entities of both short- and long-term orientation; therefore, the institutional solutions should seek a commonly accepted middle ground through mediation between opposing teams and conflicting values.

Bostrom, Dafoe, and Flynn propose a set of desirable policies that acquire a distinctive character in a super-intelligent AI world. They mention, for instance, rapid progress, AI safety, conditional stabilization, non-turbulence, universal benefit, continuity, wisdom, speed and decisiveness, adaptability, and autonomy. The success of the policies depends on understanding the workings of morality, psychology, global public opinion, culture, and religion.[2]

When implementing AI governance, there is a risk of imposing some kind of totalitarianism against the majority. To prevent such totalitarianism and a form of imposition against the majority, in various circumstances and different combinations, several states around the world have worked towards establishing transparent and fair elections, ensuring protected rights to political expression, promoting the rule of law, judicial independence, power sharing, balance of state power, constitutionally protected rights and federal systems of government.

AI governance is often coupled with safety measures to help humanity evolve beneficially alongside AI. AI safety focuses on the technical issues of the way it is built and centers on the institutions and contexts in which AI is created and used. AI governance seeks to maximize the chances that the people who create advanced artificial intelligence have of putting it to good use based on goals, motivation, worldview, time, training, and resources for the benefit of humanity.[3]

Advanced AI could play a critical role in solving global problems, such as climate change and international conflicts, and contribute substantially to improving health conditions, prosperity, and sustainable development.[4]

Examining procedural dimensions of AI governance reveals the significance of standardization and certification in determining its societal implication. There is a growing trend among standardization organizations worldwide to formulate standards for creating and deploying artificial intelligence.[5] These guidelines aim to provide uniform protocols for assuring safety, interoperability, and effective data management. The flawless interaction of AI systems from different nations is vital in international situations. The principles serve as the fundamental framework of a governance model that aims to be all-encompassing, accommodating various entities, including governmental and non-governmental players, within an intricate network of duties and anticipated outcomes.

Another fundamental component is data governance. The acquisition, storage, and retrieval of data assume critical importance as a governance concern due to the substantial data requirements for training and operating AI systems. This endeavor guarantees that data utilization adheres to legal frameworks while aligning with overarching goals such as national security, economic competitiveness, and technological innovation.

Furthermore, the advancement of artificial intelligence has stimulated the emergence of Public-Private Partnerships to regulate its implementation efficiently.[6] Public-private partnerships in artificial intelligence governance bring together the dynamic and innovative capabilities of the private sector while also including the regulatory supervision and commitment to public welfare inherent in governmental institutions. Governments are crucial in establishing legal and policy frameworks for deploying artificial intelligence. On the other hand, industry players offer their technical skills and market-driven viewpoints to develop and implement AI. The primary objective of these collaborations is to cultivate an ecosystem that combines scientific advancement with social responsibility to minimize potential hazards and maximize benefits.

Besides, there is a growing emphasis on using artificial intelligence inside the legal system. The emergence of tech startups[7] using technologies that can engage in legal reasoning and decision-making raises inquiries regarding the appropriate governance mechanisms for such systems. The crucial aspect lies in achieving a harmonious equilibrium wherein artificial intelligence may enhance the effectiveness and uniformity of legal procedures while simultaneously upholding the indispensable human qualities of empathy and ethical deliberation that are the bedrock of justice systems.

However, it is necessary to consider possible drawbacks, such as the strong totalitarianism triggered by advanced lie detection, social manipulation, autonomous weapons, and ubiquitous physical sensors and digital traces. Power and control could radically shift from the public to totalitarian elites and leaders, rendering democratic regimes vulnerable to totalitarianism, regression, and the usurpation of social goods. There is also the risk of preemptive wars due to the detection of potential threats combined with the fact that an advanced AI system could make the dynamics of a crisis more complex and unpredictable, escalating faster than humans could manage, increasing the risk of accidental war.[8]

There is a probability that when nations strive for economic growth and the development of advanced artificial intelligence, they will create superhuman AI systems that are not fully aligned with human values, leading to a diminution of the value of humans and the values that define their lives.[9] Intense competition is likely to create a systematic erosion of values, especially where there is a sense of power and wealth. Stabilization against any aberration necessitates more research and institutionalization of security measures, which require time and resources to build legitimate institutions and regulatory policies with a realistic human-centered dimension.

The role of governments is essential in creating systems of accountability and redress by using algorithmic tools to address adverse impacts that cannot be managed with immediate effectiveness by private mechanisms. A step forward is the General Data Protection Regulation (GDPR), which was adopted on May 24, 2016 and is applicable as of May 25, 2018 in the European Union, "on the protection of individuals with regard to the processing of personal data and on the free movement of such data . . . to strengthen the fundamental rights of individuals in the digital age,"[10] on the automated processing of their data. The "explanatory" algorithmic nature of the GDPR has already proven successful in several areas; however, the developments in AI create additional challenges and a need for institutionalization in a framework of cooperation between technology companies, governments, and representatives of different communities to ensure respect for human rights.[11]

The problems in governance that humanity faces today will continue to be encountered in the future with antagonism in a broader military-economic sector, in the distribution of wealth, and in safeguarding human rights. Governance needs the assistance of AI alongside a constitutional commitment and institutional mechanisms to ensure quality principles, realistic planning to level the playing field, and the proper use of AI for the benefit of citizens.

Meanwhile, there are concerns about incidents of misuse or harmful use of AI, prompting the search and finding of policies so that governance can better transition to advanced AI systems, focusing on political, economic, military, intergovernmental, and ethical dimensions with a human-centered perspective.

AI in the twenty-first century will bring opportunities and risks for humanity. Due to its potential, AI will be capable of radically transforming prosperity, wealth, or power to a degree comparable in magnitude to the nuclear or even the industrial revolution on a global scale.

## NOTES

1. Allan Dafoe, "AI Governance: A Research Agenda," Governance of AI Program Future of Humanity Institute, First Draft, July 2017, August 27, 2018 [https://www.fhi.ox.ac.uk/wp-content/uploads/GovAIAgenda.pdf(03/12/2020)].

2. Nick Bostrom, Allan Dafoe, and Carrick Flynn, "Public Policy and Superintelligent AI: A Vector Field Approach," *Future of Humanity Institute,* 2018 [http://www.nickbostrom.com/papers/aipolicy.pdf (03/12/2020)]. Future of Life, "The Asilomar AI Principles," California, January 2017 [https://futureoflife.org/ai-principles (03/12/2020)]. Dafoe, "AI Governance: A Research Agenda," p. 50.

3. Dafoe, "AI Governance: A Research Agenda," p. 6. Dario Amodei, Chris Olah, Jacob Steinhardt, Paul Christiano, John Schulman, and Dan Mané, *Concrete Problems in AI Safety* [June 21, 2016. http://arxiv.org/abs/1606.06565 (07/12/2020)]. Russell Stuart, Daniel Dewey, and Max Tegmark, "Research Priorities for Robust and Beneficial Artificial Intelligence," *Future of Life Institute—AI Magazine*, 2015, pp. 105–114 [https://futureoflife.org/data/documents/research_priorities.pdf?x90991 (07/12/2020)].

4. Nick Bostrom, Allan Dafoe, and Carrick Flynn, "Public Policy and Superintelligent AI: A Vector Field Approach," *Future of Humanity Institute*, 2018 [http://www.nickbostrom.com/papers/aipolicy.pdf (08/12/2020)].

5. Martin Bers, "Standardizing AI—The Case of the European Commission's Proposal for an Artificial Intelligence Act," *The Cambridge Handbook of Artificial Intelligence: Global Perspectives on Law and Ethics*, August 2021, p. 1 [https://ssrn.com/abstract=3900378 or http://dx.doi.org/10.2139/ssrn.3900378 (20/08/2022)].

6. A. Ballantyne and C. Stewart, "Big Data and Public-Private Partnerships in Healthcare and Research," ABR, Vol. 11, 2019, pp. 315–326 [https://doi.org/10.1007/s41649-019-00100-7 (11/09/2023)].

7. S. Praduroux (University of Turin, Italy), and V. de Paiva (Nuance Communications, Sunnyvale), L. di Caro (University of Turin, Italy), "Legal Tech Start-ups: State of the Art and Trends," p. 1 [https://vcvpaiva.github.io/includes/pubs/2016-legal.pdf (21/09/2023)].

8. Dafoe, "AI Governance: A Research Agenda," p. 7.

9. Stuart Armstrong, Nick Bostrom, and Carl Shulman, "Racing to the Precipice: A Model of Artificial Intelligence Development," *Technical Report. Future of Humanity Institute*, 2013–1, pp. 1–8 [https://www.fhi.ox.ac.uk/wp-content/uploads/Racing-to-the-precipice-a-model-of-artificial-intelligence-development.pdf(09/12/2020)]. Nick Bostrom, *Superintelligence: Paths, Dangers, Strategies*, ed. Oxford University Press (Oxford: Oxford University Press, 2014), pp. 5–15. Eliezer Yudkowsky, "Artificial Intelligence as a Positive and Negative Factor in Global Risk," in *Global Catastrophic Risks,* ed. Nick Bostrom and Cirkovic M. Milan (New York: Oxford University Press, 2008), pp. 308–345. Bruce Schneier, "Resources on Existential Risk—for: Catastrophic Risk: Technologies and Policies," *Berkman Center for Internet and Society, Harvard University*, 2015 [https://futureoflife.org/data/documents/Existential%20Risk%20Resources%20(2015-08-24).pdf?x70892(18/12/2020)].

10. European Union, "Regulation (EU) 2018/1725 of the European Parliament and of the Council," 2018 [https://ec. europa.eu/info/law/law-topic/data-protection/data-protection-eu_el 2016–2018, https://eur-lex.europa.eu/legal-content/EN/TXT/HTML/?uri=CELEX:32018R1725&from=EN (10/03/2021)].

11. Raso, Hilligoss, Krishnamurthy, Bavitz, Levin, "Artificial Intelligence & Human Rights: Opportunities and Risks," p. 57.

# Chapter 5

# Theological Approaches of AI

According to Dan Brown, a change is taking place globally, as it becomes apparent that traditional religions are going through a period of crisis, with secularization as the main reason, along with other causes. Humankind—in the author's viewpoint—is moving away from traditional religions and towards what is regarded as the "new god" of society, i.e., towards technology. Brown argues, "In times past, we as human beings always looked to God, but in the present age, we always look to smartphones, computers, and other devices."[1]

On whether a strong AI would eventually succeed religion, Prisco argues that: "It is only fair to let AI have access to the teachings of all the world's religions. Then they can choose what they want to believe," "but I think it's highly unlikely that AI would choose to believe in the petty, provincial aspects of traditional religions. At the same time, I think they would be interested in enlightened spirituality and religious cosmology, or eschatology, and develop their own versions."[2]

Another perspective is that AI tends to replace traditional religions, as it develops tools and methods to expand the human mind to achieve intellectual enlightenment and self-knowledge, like Hinduism and Buddhism.[3] With the assistance of artificial intelligence, a new form of collective consciousness is emerging, which, according to some people, will replace religion.[4] Since artificial intelligence takes care of everyday affairs, technology leaves only a limited margin for human error. Therefore, this is an extraordinary God.[5]

Super-intelligent AI or artificial general intelligence (AGI) is a powerful AI equal to or greater than human beings, creating autonomous, self-aware entities of super-intelligence. The idea of a thinking machine being able to compete with human intelligence—a machine that could quickly become much smarter than humans—also has religious implications.

Some believers want to ensure that every super-intelligent AI is aware of God. As artificial intelligence advances, religious questions and concerns arise. Presbyterian Christopher Benek stated that "if AI is autonomous, then we should encourage it to participate in Christ's redemptive purposes in the world. It's redemption to all of creation, even artificial intelligence."[6]

Martine Rothblatt believes that anything that values life or God or has the potential to value life or God has some kind of soul. She also studies religion in the time of artificial intelligence, uploads, and mind clones—software versions of human minds—and considers that mind clones are consistent with classical interpretations of religion and will be welcome in the future.[7]

Rodney Brooks argues that we are simply biological machines, and the fact that we are rational, have consciousness, and have emotions proves that machines can be reasonable because we are. Every individual is also a machine, a set of biochemical interactions that operate according to rules; therefore, machines can also interact with the world.[8]

The major Abrahamic religions—Judaism, Christianity, and Islam, representing about two-thirds of the world's population—teach about the soul based on the common texts of the Old Testament, especially Genesis.

Understanding the "soul" is essential in the upcoming era of AI engineering, as well as the issues of what exactly are the human soul and conscience and what is the cornerstone of life that DARPA, Google, and IBM are trying to explain, unveil and reproduce in purely technocratic terms.

Brantlinger states, "our theology will collapse with the advent of intelligent machines."[9] AI is used in processes that achieve the same or even better results as human intelligence in some fields. For example, Deep Blue, the chess-playing computer system, defeated World Chess Champion Gary Kasparov in 1997 (with a score of 3.5 to 2.5 in a six-game match).[10]

From artificial intelligence to whole-brain emulation, advanced technologies are heralding an ever-increasing number of miracles. For the most part, the growing techno-religious sentiment is a feeling or intuition based largely on the similarities between the promises of technology and those of religion.[11]

Although it is not new, people "transmit" their inner thoughts into "space," and technology has always shaped religious practice and faith. "Technology does feel and look and act like God, at least sometimes," says John Modern, Professor of Religious Studies at Franklin & Marshall College, "so it's certainly logical that someone would see the power of technology and locate their faith in it."[12]

William Dembski states that "without such (intelligent) machines on the horizon, I feel secure in my 'archaic' theology," referring to the achievements in the field of artificial intelligence.[13]

"Cosmic perspective does not care how it views the universe whether it is as information or quantum information or supercomputing or positive or otherwise," writes R. U. Sirius. "Nor does it ask anyone to commit to AGI [artificial general intelligence] or fusion brain uploads or computer brain interfaces as the best means of moving forward. Instead, it seeks to enrich the human universe with a posture of joy, growth, and open-minded choices. Secularism believes that science, as well as religion and philosophy, in their current forms, may prove too limited for the task of understanding life, mind, society and reality."[14]

A growing number of scientists, including Hans Peter Moravec, Allen Newell, Ray Kurzweil, and Hugo de Garis,[15] openly describe technological progress in religious terms.

The emergence of this techno-religious sentiment was put forward by the twentieth-century philosopher Ellul Jacques, who, in his 1954 work *The Technological Society*, noted that in the face of technological progress, man creates for himself a new religion of rational and technical order, justifies his work, and is justified by it.[16] Historian Yuval Harari, in his book *Homo Deus: A Brief History of Tomorrow*, qualifies the belief in the power of data and the promise of artificial intelligence as a religious movement that Harari calls Dataism. In other words, he argues that, just as religious mythologies legitimized divine power, so too are high-tech gurus and Silicon Valley prophets creating a new universal narrative that legitimizes the power of algorithms and Big Data.[17]

Mike McHargue, the author of *Finding God in the Waves: How I Lost My Faith and Found It Again through Science*,[18] believes that the rise of AI will remove any ambiguities in the ways that many Christians and other Abrahamic religions have defined terms like "conscience" and "soul." Based on a syllogistic association, McHargue poses the question, if you have a soul and create a physical copy of yourself, do you assume that your physical copy also has a soul?," and "If we learn to encode a human brain digitally, then AI would be a digital version of ourselves. If you create a digital copy, does your digital copy also have a soul?"[19]

## NOTES

1. CBS News, *The Evolution of Dan Brown*, 2017 [https://www.cbsnews.com/video/theevolution-of-dan-brown/ (19/01/2021)].

2. Wesley Smith, "Humans Will Evolve Beyond Ourselves via Technology? At Least the Idea Offers Some Comic Relief," *Evolutionnews*, March 3, 2015 [https://evolutionnews.org/2015/03/humans _will_evo/ (03/02/2021)].

3. Melissa Rendsburg, *The Impact of Artificial Intelligence on Religion: Reconciling a New Relationship with God*, July 2019, pp. 17, 18 [https://www.academia.edu/40235300/The_Impact_of_Artificial_Intelligence_on_Religion_Reconciling_a_New_Relationship_with_ God (17/01/2021)].

4. Rendsburg, *The Impact of Artificial Intelligence on Religion: Reconciling a New Relationship with God*, p. 18.

5. Helena Bassil-Morozow, "God Is an Algorithm: Why We're Closer to a Black Mirror–Style Reality than We Think," *The Conversation*, January 31, 2018 [https://theconversation.com/god-is-an-algorithm-why-were-closer-to-a-black-mirror-style-reality-than-we-think-90669 (01/02/2021)].

6. Christopher Benek, "Thinking West: Robots Taking Selfies," *El Portal Literary Journal*, February 25, 2015 [https://elportaljournal.com/category/weird-s// (05/03/2021)]. Zoltan Istvan, "When Superintelligent AI Arrives, Will Religions Try to Convert It?" *Gizmodo*, 2/04/15 [https://gizmodo.com/when-superintelligent-ai-arrives-will-religions-try-t-1682837922/ (03/04/2021)].

7. Martine Rothblatt, *Virtually Human: The Promise—and the Peril—of Digital Immortality*, (New York: St. Martin's Press, 2014), pp. 283–295.

8. Rodney Allen Brooks, *Flesh and Machine Robots Will Change Us*, (New York: Pantheon, 2002), p. 175.

9. William Dembski, "Conflating Matter and Mind," *Perspectives on Science and Christian Faith* 43, no. 2 (June 1991): 107–111.

10. Russell C. Bjork, *Artificial Intelligence and the Soul*, 2008, p. 196 [https://www.asa3.org/ASA/PSCF/2008/PSCF6-08Bjork.pdf (02/02/20210].

11. Tonar Remington, *The Church of Artificial Intelligence: A Religion in Need of a Responsible Theology*, Aug. 2017 [https://medium.com/@AItheist/the-church-of-artificial-intelligence-a-religion-in-search-of-a-theology-fa7e73f46900 (12/12/2020)].

12. Jessica Roy, "The Rapture of the Nerds." John Modern, *Neuromatic; or, a Particular History of Religion and the Brain*, Dec. 2020 [http://www.johnmodern.com/2020/12/15/neuromatic-or-a-particular-history-of-religion-and-the-brain/ (05/03/2021)].

13. Dembski, "Conflating Matter and Mind," p. 109.

14. Jay Cornell, R. U. Sirius, *Transcendence: The Disinformation Encyclopedia of Transhumanism and the Singularity Paperback* (New York: Disinformation Company Ltd, 2015), pp. 21–42.

15. Remington, *The Church of Artificial Intelligence: A Religion in Need of a Responsible Theology*.

16. Ellul Jacques, ed., *The Technological Society* (New York: Vintage Books, 1964), p. 324.

17. Yuval Harari, ed., *Homo Deus: A Brief History of Tomorrow* (New York: Harper, 2017), pp. 15–25.

18. Mike McHargue, ed., *Finding God in the Waves: How I Lost My Faith and Found It Again through Science* (Colorado: Convergent Books, 2016).

19. McHargue, *Finding God in the Waves: How I Lost My Faith and Found It Again through Science*, pp. 15–30. Cellan-Jones Rory, "Stephen Hawking Warns Artificial Intelligence Could End Mankind," *BBC*, December 2, 2014 [https://www.bbc.com/news/technology-30290540 (20/02/2021)].

## *Chapter 6*

# Approaches to the Concepts of Salvation and Immortality through the Uses of AI

Victoria Nelson, in her work *The Secret Life of Puppets*, points out that modern Americans maintain a subconscious belief in the divinity of machines, as the sense of the supernatural and the longing for immortality has shifted from religion to such expressions of popular culture as superheroes, robots, and cyborgs, and that modern civilization has suppressed people's spiritual instincts.[1] Employing science fiction as evidence, Nelson argues that artificial humans represent a combination of God, an externalized soul, and the Divine Human. She holds that "the attribution of near omnipotence to machines proves their divine potential. However, power alone is not sufficient for the definition of divinity."[2]

At this point, we must recognize the rich tapestry of non-theistic religious and philosophical perspectives on the role of AI in the universe. These belief systems, which often emphasize the interconnectedness of all things and the power of reason, offer unique insights into how AI can be understood and incorporated into our lives.

In contrast, the concept of the divine varies across religious traditions. Still, it generally refers to a transcendent, ultimate reality or supreme being responsible for the universe's creation, sustenance, and governance.

From a theistic standpoint, AI can be seen as a product of divine will, either directly or indirectly. In monotheistic religions like Christianity, Islam, and Judaism, God is often considered the ultimate source of all knowledge and wisdom. In this view, human Intelligence and creativity, which have given rise to AI, are gifts from God, and therefore, AI can be seen as a divine creation.

In polytheistic religions like Hinduism, various gods and goddesses are associated with knowledge, wisdom, and creativity. The development of AI could be attributed to the influence of these divine beings, who bless humans to create such advanced technologies. From this perspective, the emergence of AI serves as a reminder of the divine origin of human Intelligence and creativity. It also carries a sense of responsibility, as humans are entrusted with the proper use of AI under divine guidance and ethical principles.

Another possibility is that advanced AI systems could be manifestations of the divine. Some futurists and theologians have speculated that AI could eventually reach a level of sophistication and consciousness that would qualify it as a new form of heavenly or spiritual being. This idea is closely related to the "Technological Singularity" concept, in which AI surpasses human Intelligence and begins to evolve, potentially developing its sense of self and consciousness.

In Rudolph Otto's references to the divine realm and religion, God intimidates and terrifies human beings. At the same time, the Holy holds a fascination because they are the source of the ultimate good of salvation. The luminous being (or object) offers salvation to humans, who must find ways to relax and enjoy themselves. From this point of view, religion becomes a technology of the Good and serves as the most valuable thing in human faith.[3]

According to Otto, the Holy provides love and salvation and, at the same time, overwhelms and terrifies us with power and possible wrath. In attempting to describe the fundamental nature of religious experience, Otto presents us with a concurrence of contrasts: the mystery of horror and light with love. These contradictions also characterize the technology of the twenty-first century, which both terrifies us as inhuman and extinction-threatened and fascinates us. Many believe, as it seems, that it entices us with promises of "salvation" in the form of a "casual" return to Eden.[4]

Thus, technology has long been associated with cosmic salvation and the Kingdom of God in the Euro-American, mainly Protestant Christianity. In *The Religion of Technology*, David Noble believes that artificial Intelligence is seen as a step towards a final revolution and confrontation between creator and creation, which is concluded through the return of the mind to its divine origin. He tracks the millenarian expectations that led medieval and modern Christians to believe God intended to create a second Eden.[5] The hope for a second Eden promises a life of recreation, perhaps even immortality. Meanwhile, intelligent machines are always on the verge of rebellion and are taking over the planet; according to eschatological scenarios, robots attack humans, possibly enslaving them. The portrayal of robots in science fiction suggests that the fear of technological wrath goes hand in hand with the hope for a new Eden.[6]

Perhaps this is why Patrick Brantlinger described science fiction as "a form of apocalyptic fantasy verging on religious myth." In science fiction, robots threaten humanity with displacement, from the economy to their lives. Yet, they can never detach themselves from the promise of a better life by robotic technology. Furthermore, although there may be some correlation between technology and the Holy in science fiction, modern human beings do not seem to accept or respect it wholeheartedly. Thus, what can be attributed to modern technology and science fiction is ambiguous.[7]

Also, popular scientific publications in robotics and AI reveal a striking amalgamation between apocalyptic religious thought and scientific research. The three main characteristics of the early Jewish and Christian apocalyptic tradition are alienation from the world, a desire to create a heavenly new world, and the transformation of human beings to live in this new world in cleansed bodies. In "Apocalyptic AI," these characteristics are attributed to scientific authority. Proponents of apocalyptic artificial Intelligence, frustrated by the limitations of physical life, are looking forward to a virtual world inhabited by intelligent machines and human beings who have left their bodies. Having uploaded their consciousness into machines, human beings will possess enhanced mental abilities and, through their unrestricted replication, immortality.[8]

Moravec and Kurzweil argue that robotic technology can endow us with personal immortality. In this context, the objective is not so much to create independent, intelligent entities as to produce virtual brains in which a person's personality can be "uploaded," which allegedly will allow, in combination with the creation of backups (periodically), to make an individual immune to death by accident, disease, or the ravages of old age.[9] They assert this because religious beliefs are essential for attaining rewards such as eternal life.[10]

According to Kurzweil, it is promising that humans can achieve immortality by uploading their minds into advanced AI systems. Therefore, uploading the human mind into an artificial intelligence system could undoubtedly "answer" the "soul" question, according to the research of Google's engineer, Ray Kurzweil. Kurzweil argues in *Time* that: "People have invested a lot of personal effort in certain philosophies dealing with the issue of life and death. I mean, that's the main reason we have religion."[11]

Since God's plan for humanity is redemption and eternal salvation, it is suggested by some people that AI can be regarded as an alternative to hope. For instance, Hans Moravec[12] and Ray Kurzweil claim that in the not-too-distant future, we will see intelligent machines whose mental capabilities far exceed the biological abilities of humans and whose powers will make it possible to solve problems that have long plagued humanity.[13]

Martine says, "It's for us that God creates our collective efforts to make technology more ubiquitous, omnipotent, and moral," and "when we can joyfully all experience techno-immortality, then God is complete." When these standpoints are received with respect and faith in the power of science, a religious belief also begins to be reflected. Especially when the capabilities that future technologies are believed to have, such as the omnipotence of technology and the ability to raise the dead, it is obvious that the believers of these theories attribute AI to a character reminiscent of the overall religious principles referring to God. Thus, technology tends to take on the characteristics of a religion.[14]

Robert Geraci suggests in his 2010 book, *Apocalyptic AI: Visions of Heaven in Robotics, Artificial Intelligence, and Virtual Reality,* that AI can fulfill the same role in apocalyptic fantasies as a single divine entity. Bearing in mind that biblical revelation is an optimistic cosmic transformation, he also draws parallels with the goals of artificial Intelligence, which often describe hopeful expectations for a world to come, i.e., an AI eschatology. In an early part of this work, Geraci builds on Rudolph Otto's account of the divine (1917) and speaks of the divine as a tremendous mystery and fascinating with a sense of awe: "mysterium tremendum et fascinans." Geraci uses Otto's definitions of God and religion to identify an analogous type of awe and terror and to indicate that the terrifying at different moments in our history can be God or, in our modern contemporary world, artificial intelligence. Other times, Geraci's work has engaged with virtual worlds, drawing attention to the part of transhumanists, such as Giulio Prisco, mentioned later, in claiming new possible settings for the practice and evolution of religion towards transhumanist ends. In such settings, including *Second Life* and *World of Warcraft* (the MMORPG—a massively multiplayer online role-playing game), Geraci argues that we take a step closer to fulfilling transhuman salvation—"a heavenly kingdom to inhabit."[15]

Another parameter that arises is the potential for AI to provide definitive answers to our deep questions about life, death, and the afterlife. In this perspective, individuals might be drawn to the idea of existential continuity through AI, contemplating the upload of their consciousness online. An action that could serve as a means for descendants to interact with their ancestors, seeking guidance on everyday dilemmas: "What would you do in this situation? How should I act at work when faced with this issue from my manager?"

These possible scenarios challenge conventional religious notions of the soul and afterlife, presenting a form of spiritual fulfillment that's hard to counter. Such concepts could gain traction, especially if promoted by spiritual leaders through modern platforms like social media or depicted in science fiction narratives where protagonists grapple with the decision of consciousness continuation.

Films that question the essence of life and the thin line between humanity and advanced technology are not novel. Classics like *2001: A Space Odyssey* (1968) and *Blade Runner* (1982) were pioneers in exploring the concept of sentient machines and the potential perils technology poses to humanity.

## NOTES

1. Victoria Nelson, ed., *The Secret Life of Puppets* (Cambridge: Harvard University Press, 2001), pp. 240–251.

2. Nelson, *The Secret Life of Puppets*, p. 269.

3. Rudolph Otto, *The Idea of the Holy*, trans. J. W. Harvey (London: Oxford [1917] 1923), pp. 33–35.

4. Otto, *The Idea of the Holy*, pp. 30–35. R. Geraci, "Robots and the Sacred in Science and Science Fiction: Theological Implications of Artificial Intelligence," *Journal of the American Academy of Religion* 76, no. 1 (2008): 965.

5. David Noble, ed., *The Religion of Technology: The Divinity of Man and the Spirit of Invention* (New York: Knopf Doubleday Publishing Group, 2013), pp. 224–270. Geraci, "Robots and the Sacred in Science and Science Fiction: Theological Implications of Artificial Intelligence," p. 965.

6. Geraci, "Robots and the Sacred in Science and Science Fiction: Theological Implications of Artificial Intelligence," p. 966.

7. Geraci, "Robots and the Sacred in Science and Science Fiction: Theological Implications of Artificial Intelligence," p. 966.

8. Geraci, "Apocalyptic AI: Religion and the Promise of Artificial Intelligence," p. 138.

9. Ray Kurzweil, *The Singularity Is Near* (New York: Viking, 2005), pp. 198–203.

10. Rodney Stark and William Sims Bainbridge, *A Theory of Religion*, (New York: Peter Lang, 1987), p. 340.

11. Lev Grossman, "2045: The Year Man Becomes Immortal," *Time*, Feb. 10, 2011 [http://content.time.com/time/magazine/article/0,9171,2048299–4,00.html (03/01/2021)]. Jason Summers, "Can Singularity and Christianity Coexist?"

12. Moravec, a robotics researcher, a "Living Legend" and the founder of Apocalyptic AI, made significant contributions to robot vision and navigation and helped design the Stanford Cart, the first autonomous computer-controlled vehicle. Lee Gutkind, ed., *Almost Human: Making Robots Think* (New York: W.W. Norton, 2006). Geraci, "Apocalyptic AI: Religion and the Promise of Artificial Intelligence," p. 139.

13. Hans Moravec, *Robot: Mere Machine to Transcendent Mind* (New York: Oxford University Press, 1999). Ray Kurzweil, ed., *The Age of Spiritual Machines* (New York: Viking, 1999), pp. 125–137. Kurzweil, *The Singularity Is Near*, pp. 14–29.

14. Roy, "The Rapture of the Nerds."

15. Robert Geraci, *Virtually Sacred: Myth and Meaning in* World of Warcraft *and* Second Life (Oxford: Oxford University Press, 2014), p. 177.

# Chapter 7

# The "Apocalyptic" Artificial Intelligence

In the late twentieth and early twenty-first centuries, a body of literature on the relationship between religion and science in modern life was produced by AI pioneers. Prominent figures of this movement are the roboticist Hans Moravec, the AI inventor Ray Kurzweil, the "builder" of neural networks Hugo de Garis, and the British roboticist Kevin Warwick. Hans Moravec and Ray Kurzweil have been widely recognized for their scientific contributions, as well as for their achievements in the field of the so-called apocalyptic artificial intelligence (AAI).

The authors predict the course of history revolving around robotics and AI technology for the next fifty years and explore the transcendent realm of cyberspace. Ray Kurzweil identifies an "angelic figure" from the transcendent future era who offers advice and interpretation.[1]

According to the theories presented by the authors mentioned above, in ancient apocalyptic texts and spiritual movements, the one God acts to guarantee the victory of good over the forces of evil. In AAI, evolution is a safeguard for the triumph of intelligent calculation over the forces of ignorance and inefficiency, providing transcendent guarantees for the future.[2]

Through the dualistic worldview of a world divided into good and evil, AAI embodies long-standing religious dreams of purity, perfection, and immortality that can be realized through scientific and technological achievements.[3] Also, according to these theories, AAI allegedly assumes a role in mind-body dualism, in the attempted alienation and transcendence of the world and the body, thereby interpreting ancient apocalyptic approaches on this basis.[4]

Several AI and robotics pioneers speak the language of the apocalyptic. Proponents predict a radical chasm in history that is bridged through their modern theory of alienation. This prediction necessitates the creation of a new world where mechanical life replaces biological life.

According to these viewpoints, human beings will reject the limitations of their bodies in favor of mechanical substitutes and virtual bodies that will live forever in eternal bliss.[5]

The distinctive elements that characterize the Jewish and Christian Revelation are the alienation of the world of corruption, the longing for the creation of a heavenly new world, and the transformation of people to inhabit this world in purified bodies and souls. Scientific publications in robotics and AI reveal a striking merge between apocalyptic religious thought and scientific research. The characteristics above are attributed, according to AAI, to scientific authority. AI advocates are looking towards a virtual world inhabited by intelligent machines and humans who have left their bodies without limitations. Having uploaded their consciousness into machines, humans will possess enhanced mental abilities and immortality through unbounded replication.[6]

For the followers of apocalyptic AI, the human body has a few major limitations relating to its limited lifespan. Moreover, a mind trapped in a human body can only learn with effort, think slowly, and struggle to acquire and transmit knowledge. "Protein-based life forms," say the proponents of apocalyptic artificial intelligence, "will never be able to think like machines in the future."[7]

It is noticeable that the intersection of apocalyptic AI and ancient spiritual philosophies exemplifies the combination of humanity's primordial aspirations and the rapid pace of technological advances. Through the technological lens, this collision of domains reveals a future in which humans are not merely surviving but evolving into superior beings—a type of Intelligence plus sentient beings.

Ray Kurzweil predicted that Singularity is near and will happen sometime in the next thirty years.[8] In contrast, Paul Allen and Mark Greaves in the MIT Technology Review present a different angle: "While we suppose this kind of Singularity might one day occur, we don't think it is near. We believe it will be a very long time coming."[9]

One can draw a parallel between the utopias depicted in religious texts and those promised by AI futurists. The human consciousness is wired to achieve freedom from everyday life's cage by completing Herculean labors and overcoming mortality.

Throughout history, humanity has been preoccupied with the concept of transcending limitations. The Tower of Babel, Icarus' wings, and even Ponce de León's search for the Fountain of Youth exemplify this never-ending quest to surpass human limitations. Today, as we are on the verge of realizing a virtual world beyond our mortal existence, one is reminded of ancient stories in which heroes ventured into mystical regions in search of knowledge or power. A rekindling of these narratives is taking place.

The tales told by proponents of apocalyptic AI, such as Moravec and Kurzweil, resonate with these timeless tales. In the utopia driven by artificial Intelligence, humanity does not seek the favor of celestial entities but instead channels machine intelligence's power. The Garden of Eden is no longer a verdant expanse but a boundless, imperfect-free digital landscape.

While the apocalyptic AI movement portrays a rosy picture of the future, full of promises of never-ending happiness and limitless knowledge, it is imperative to proceed cautiously. In the same way that Icarus was undone by his hubris, humanity must ensure that its voyage towards a machine-driven utopia does not result in an unanticipated descent.

Considering these compelling prospects of a technologically enhanced afterlife, it is worthwhile to ponder: Are we on the verge of realizing paradise through our machines, or is this just another chapter in humanity's never-ending search for the elusive elixir of immortality? Time, or possibly our intelligent machines, will inform.

As mentioned, Kurzweil believes that in 2045, mechanical bodies will greatly enhance the powers of the human mind, as nanotechnology will allow human beings to build far superior biological bodies equipped with flexibility, self-healing, resilience to environmental changes, and the ability to compute faster. It will be possible to transfer consciousness and the human mind and eliminate the needs arising from the connection to real (physical) bodies. Since there will be a virtual environment, the virtual body will "replace" the real body.[10] Consequently, identification with biological data, which virtual bodies will replace, will cease. Thus, in the future, human beings will reconfigure their bodies to participate in the Kingdom, either as cyborgs, robots, or software, thereby eradicating pain, participating in a global knowledge network, and living forever in the Virtual Kingdom.[11]

The Virtual Kingdom discards both traditional humanity and traditional religion. The absolute value of a human, life, as a logical calculation must be liberated from the body and possibly the emotions that characterize life as we know it. The apocalyptic writers claim that humans will return to the undifferentiated wholeness of good[12] by eliminating the physical and embracing the virtual.

The speeds achievable in the human brain cannot compare with those of computers, whose silicon substrate is much more efficient at transmitting information. Limited memory and insufficient precision hinder human brains so these problems will be overcome with the transition to mechanical life.[13]

Slow computation, limited recall, insufficient capability to communicate information, and death all restrict the mind from reaching its full potential, which can radically shift its life. The revelation of AI has its basis because people fear dying or knowledge loss, and evolutionary history inevitably moves in this direction.[14]

According to the AAI, the value of a living person comes from the knowledge they possess rather than being innate to life or relying on social relations. The artificial intelligence apocalypse will end the "involuntary loss of knowledge and function, which is the worst aspect of personal death," as the death of knowledge matters more than the death of people.[15]

The fear of death is quite common, but proponents of apocalyptic artificial Intelligence view the traditional and widely held belief in souls and spirits as a weak psychological trick. According to AI pioneer Marvin Minsky, "all" is an insinuation that we are powerless to improve ourselves.[16] It is the loss of knowledge that cannot be overcome. Religion, according to AAI, can be dealt with through technology. Minsky's approach expresses his sorrow for the religious failures to understand human life. Still, his contrasting exuberance for the possibilities of artificial Intelligence stems from the fact that AAI proclaims solutions for the future with the same enthusiasm that can be found in ancient apocalypticism.[17]

Similar to the way that many early Jews and Christians believed that the intervention of God in history was imminent, modern approaches to AAI also foresee that, at some point, a pivotal change will occur. The gradual change is limited to apocalyptic visions, whereas Apocalypse devotees expect a sudden revolution.[18]

Michael Benedikt argues that "cyberspace opens the doors to the Heavenly City of Revelations." Architecture, in its utopian manifestations, seeks to bring about the same conditions it depicts, combining art and technology to reveal the Kingdom. Thus, cyberspace will offer human beings unlimited joy through an idyllic environment and infinite personal experience. The eschatological architectural imagination of the Kingdom of Benedict displays the connection between virtual reality and Christian salvation.[19] According to Benedikt, a good life in an egalitarian society will be guaranteed in cyberspace by eliminating necessity and happiness,[20] whereas the Virtual Kingdom will "guarantee" immortality.[21]

According to Kurzweil and Moravec, mainstream religious belief teaches that after the end of history, God will create a new world and resurrect humanity in glorified new bodies to enjoy this world eternally. Proponents of apocalyptic artificial Intelligence cannot rely on divine powers to guarantee the forthcoming Kingdom, so they turn to technological evolution as a transcendent guarantee for the new world.[22] Even without God, development ensures the advent of the Kingdom. AAI envisions a mechanical future in which human beings will upload their minds to machines and enjoy a virtual reality paradise with perfect virtual bodies.[23]

The AAI community believes that machines' creativity will lead to salvation. The most important task for humankind is creating artificial life where robots will accomplish what biological life cannot. "Singularity" is the point at which machines become sufficiently intelligent to begin to learn. When this starts to happen, the pace of development will be rapid, which means that in a very short time, the world will irrevocably shift from biological to mechanical.[24]

In AAI, this major event is called "Singularity," which marks a radical gap between this world and the next, a mechanical world, a Virtual Kingdom in cyberspace. According to Geraci, technological evolution with high-speed computing foreshadows the end of history and the beginning of the new world.[25]

The mechanical age will allow the establishment of a new Kingdom, which will not rely on God but will guarantee the same promises with more traditional revelations. The journey towards the Virtual Kingdom will be carried on through an Edenic earthly life before the final transcendence of mind over matter. Artificial intelligence and robotics will relieve humankind of its burdens.[26]

In this context, Kurzweil envisions a future in which necessity is an "odd idea"; no one will have to work for their daily bread, but rather, it will fall from the sky, providing free time for spiritual discoveries.[27]

Before the eventual transition to the Virtual Kingdom, Moravec holds a fictional vision of human bliss, in which robots will cause humankind to forsake the necessity for material things in a paradise that will never be lost.[28]

Moravec's "Age of Robots" is replaced by the "Age of Mind," in which the expansion of machines makes room for a "finer world." Heaven on Earth will be but a brief interlude in the historical path toward a transcendental Virtual Kingdom[29] that will cure the afflictions of the present world.

Moreover, the literature on AAI promises an escape from the time when highly intelligent machines used cyberspace to escape from the present historical reality. In Mind Fire, entire universes will be created through massive computer simulations.

History will be encapsulated and recreated in a new virtual reality.[30] All of history will coexist at an eternal moment, where the Virtual Kingdom will transcend earthly life and will be apocalyptic beyond a utopian eschatology.[31] This new AAI world is transcendental, given that it exceeds human life and replaces it with the Virtual Kingdom in a transcendental cyberspace level where pain ceases to exist, and history ends in bliss.[32]

Consequently, AAI advocates do not rely on divine powers for the forthcoming Kingdom, a godless Kingdom based on information and knowledge. In this mechanical future, human beings will upload their minds into machines and enjoy a virtual reality paradise with perfect virtual bodies.[33]

## NOTES

1. Geraci, "Apocalyptic AI: Religion and the Promise of Artificial Intelligence," p. 146.
2. Geraci, "Apocalyptic AI: Religion and the Promise of Artificial Intelligence," p. 159.
3. Geraci, "Apocalyptic AI: Religion and the Promise of Artificial Intelligence," p. 160.
4. Geraci, "Apocalyptic AI: Religion and the Promise of Artificial Intelligence," p. 161.
5. Joseph John Collins, *The Apocalyptic Imagination: An Introduction to the Jewish Matrix of Christianity* (New York: Crossroad, 1984), p. 5.
6. Geraci, "Apocalyptic AI: Religion and the Promise of Artificial Intelligence," p. 138.
7. Hans Moravec, ed., *Mind Children: The Future of Robot and Human Intelligence* (Cambridge, MA: Harvard University Press, 1988), pp. 55, 56. Hans Moravec, *Robot: Mere Machine to Transcendent Mind* (New York: Oxford University Press, 1999), p. 55. Kurzweil, *The Age of Spiritual Machines*, p. 4. Kevin Warwick, ed., *March of the Machines: The Breakthrough in Artificial Intelligence* (Chicago: University of Illinois Press, [1997] 2004), p. 178. Hugo De Garis, ed., *The Artilect War: Cosmists vs. Terrans: A Bitter Controversy Concerning Whether Humanity Should Build Godlike Massively Intelligent Machines* (Palm Springs, CA: ETC Publications, 2005), p. 103.
8. Christianna Reedy, "Kurzweil Claims That the Singularity Will Happen by 2045," Futurism, updated October 16, 2017 [https://futurism.com/kurzweil-claims-that-the-singularity-will-happen-by-2045 (20/08/2023)].
9. P. G. Allen and M. Greaves, "Paul Allen: The Singularity Isn't Near," MIT Technology Review, October 12, 2011 [https://www.technologyreview.com/2011/10/12/190773/paul-allen-the-singularity-isnt-near/ (15/08/2023)].
10. Kurzweil, *The Age of Spiritual Machines*, p. 141, 142.
11. Geraci, "Apocalyptic AI: Religion and the Promise of Artificial Intelligence," p. 154.
12. Geraci, "Apocalyptic AI: Religion and the Promise of Artificial Intelligence," p. 160.
13. Geraci, "Apocalyptic AI: Religion and the Promise of Artificial Intelligence," p. 17.
14. Geraci, "Apocalyptic AI: Religion and the Promise of Artificial Intelligence," p. 148.
15. Hans Moravec, *Mind Children: The Future of Robot and Human Intelligence*, p. 121.
16. Marvin Minsky, *Society of Mind* (New York: Simon and Schuster, 1985), p. 41.
17. Minsky, *Society of Mind*, p. 41. Geraci, "Apocalyptic AI: Religion and the Promise of Artificial Intelligence," p. 148.
18. Rennie Schoepflin, ed., *Apocalypticism in the Modern World and the Contemporary Age* (New York: Continuum, 2000), p. 428.

19. Michael Benedikt, *Introduction. In Cyberspace: First Steps* (Cambridge, MA: M.I.T. Press, 1994), p. 14.

20. Moravec, *Robot: Mere Machine to Transcendent Mind*, p. 137. Kurzweil, *The Age of Spiritual Machines*, pp. 128, 206, 236, 248. Allucquere Rosanne Stone, ed., *Will the Real Body Please Stand Up? Boundary Stories about Virtual Cultures* (Cambridge, MA: MIT Press, 1991), pp. 81–118.

21. Geraci, "Apocalyptic AI: Religion and the Promise of Artificial Intelligence," p. 149.

22. Moravec, *Robot: Mere Machine to Transcendent Mind*, p. 165. Moravec, *Mind Children: The Future of Robot and Human Intelligence*, p. 158, 159. Kurzweil, *The Age of Spiritual Machines*, p. 255. Kurzweil, *The Singularity Is Near*, p. 175.

23. Geraci, "Apocalyptic AI: Religion and the Promise of Artificial Intelligence," p. 140.

24. Kurzweil, *The Age of Spiritual Machines*, p. 248.De Garis, *The Artilect War: Cosmists vs. Terrans: A Bitter Controversy Concerning Whether Humanity Should Build Godlike Massively Intelligent Machines*, p. 158.

25. Geraci, "Apocalyptic AI: Religion and the Promise of Artificial Intelligence," p. 148.

26. Geraci, "Apocalyptic AI: Religion and the Promise of Artificial Intelligence," p. 150.

27. Kurzweil, *The Age of Spiritual Machines*, p. 249.

28. Moravec, *Robot: Mere Machine to Transcendent Mind*, p. 128.

29. Moravec, *Mind Children: The Future of Robot and Human Intelligence*, pp. 116, 163.

30. Hans Moravec, "Letter from Moravec to Penrose," in *Thinking Robots, an Aware Internet, and Cyberpunk Librarians: The 1992 LITA President's Program,* ed. R. Bruce Miller and Milton T. Wolf, *Library and Information Technology Association* (Chicago: 1992a), pp. 51–58. Moravec, *Robot: Mere Machine to Transcendent Mind*, pp. 164, 167.

31. John Joseph Collins, ed., "From Prophecy to Apocalypticism: The Expectation of the End," in *The Encyclopedia of Apocalypticism,* Volume I: *The Origins of Apocalypticism in Judaism and Christianity* (New York: Continuum Press, 2000b), pp. 129 161.

32. Geraci, "Apocalyptic AI: Religion and the Promise of Artificial Intelligence," p. 152.

33. Geraci, "Apocalyptic AI: Religion and the Promise of Artificial Intelligence," p. 140.

## *Chapter 8*

# The "Churches" of Artificial Intelligence

A significant emerging trend of the last century, rooted in the Scientific Enlightenment, is the proliferation of communication options via the Internet and mobile telephony, creating an interconnectedness of the spiritual beliefs of much of the world's population. Technological progress tends to engage the world population in a cosmic super-humanism over time and with the upbringing of new generations.

The Turing Test in 1950, since its inception, and the virtual state of *Blade Runner*, *Her,* and *ExMachina* clearly indicate humanity's fascination with the idea of machines exhibiting human-level intelligence. The Turing Test, however, was never intended to prove that machines are as intelligent as humans. It was designed to demonstrate how effectively a machine can interact with a human in a closed-ended conversation. According to Levchuk, our science's ability to create a "soul" of silicon would be an undeniable triumph of the Scientific Revolution in the "imitation game" of humans.[1]

Researchers' metaphysical questions regarding faith and artificial intelligence are reduced to whether AI has a soul. Can it both be saved and saved? "The question is if humans can be saved, then there is no reason why the thoughts and feelings of AI cannot be saved. As soon as there is a human-like AI, they will be entities just like us," argues futurist Giulio Prisco, founder of the transhumanist Turing "church."[2]

Before analyzing Turing's "church," it is necessary to establish the conditions for possessing a soul. Defining these criteria is essential to consider whether AI could possess a soul. Derived from the diverse theological traditions examined above, certain shared criteria for preserving a soul may encompass the following:

- Consciousness and self-awareness refer to the capacity to perceive and reflect upon one's thoughts, emotions, and experiences.

- Moral agency refers to making ethical decisions and being responsible for one's actions.
- Free will is the capacity to make decisions and choices without being influenced by deterministic circumstances or external forces.
- Immaterial essence refers to a non-physical and intangible essence beyond the physical body and its functions.

Arguments for AI having a soul.

The emergence of consciousness: Some philosophers and scientists argue that AI systems could give rise to subjective states of consciousness and self-awareness, indicating a non-physical essence or soul as they become more complex and human-like.

The continuity of intelligence: Proponents of the idea that AI could possess a soul argue that there may be a continuity of intelligence between human beings and advanced AI systems, suggesting that the same non-physical essence that animates human intelligence could also be present in AI.

Arguments against AI having a soul.

The materialist perspective: Materialists contend that consciousness and self-awareness are products of physical processes within the brain and that since AI systems are fundamentally different from biological organisms, they cannot possess a non-physical essence or soul.

The divine creation argument: Many religious traditions maintain that the soul is a divine creation unique to human beings and that AI, as a product of human ingenuity, cannot possess a soul.

In that light, Turing's "church" comprises seekers at the crossroads of science and religion, spirituality and technology, engineering and science fiction, mind and matter. It's about hacking religion, enlightening science, and awakening technology.[3]

The Turing "church" started as a working group bridging science and religion and has recently evolved into an online, open-source church centered around secular principles of space expansion, growth, and universal love. Similarly, Rothblatt established Terasem, a scientific "trans-religion" comparable to the Turing "church" in its scope and methodology, embarking on pioneering thought experiments.

Transhumanism, which emerged nearly a century ago and gained significant traction in the 1990s, advocates for the ethical application of technology to transcend biological limitations and augment human physical and mental capabilities. Technologies like Google Glass, artificial limbs, and even birth control exemplify how technology can enhance human biology. While most transhumanists identify as atheists or agnostics, some are drawn to new religions like Terasem, which fulfill a spiritual void for those deeply immersed in technology.[4]

Terasem's adherents are committed to exploring and promoting "personal consciousness in cyberspace"—creating intelligent mind-files. They believe that by meticulously recording thoughts and emotions, a digital replica of each person can be constructed for future use. This involves transcending the physical body and transferring the brain or personal consciousness into digital forms. These mind-files are preserved on servers using Terasem's services, with a commitment to long-term safeguarding. The vision is that these files, organized by AI software with a consciousness-oriented approach, could be integrated into an artificial body years or centuries later.[5]

Some Terasem believers are motivated by a desire similar to that shared by the followers of the most popular religions: the longing to be reunited with people who have died. Linda Chamberlain, co-founder of the cryonics company Alcor Life Extension Foundation and an active Terasemian, anticipates that one day in the future, she will be reunited with her husband Fred, who passed away a few years ago, and they will be able to explore space together. Giulio Prisco, a physicist of Terasem, says he hopes to be reunited with his mother eventually. Some of Terasem's ideas resonate with Silicon Valley's views and pioneering ideas, where millions of dollars are spent exploring how technology can change the end of life and beyond. Individuals such as Google's Larry Page and PayPal co-founder Peter Thiel are investing in projects focused on life extension and rejuvenation.[6]

Terasem is a powerful symbol of a modern way of life, where the digital and emotional worlds are increasingly intertwined. It is also a sign of the ever-evolving relationship between technology and faith. For instance, the number of Americans who call themselves "religious" has declined, even though many still describe themselves as "spiritual."

Gabriel Rothblatt, a pastor in Terasem, Florida, regards the satellite dishes in his facility as not intended for television. Their objective is to transmit the minds, memories, thoughts, and emotions of people who wish to create digital copies of themselves and launch them into space, assuming they will eventually reach some benevolent extraterrestrial beings. It's a new type of religion that seeks answers to age-old questions, driven by faith in the transformative power of technology. The new supra-religion is based on four main principles—"life is purposeful, death is optional, God is technological, and love is essential." Terasem is a "trans-religion," which means that one does not have to be a Christian or a Jew or a Muslim. Many believers embrace traditional positions held by mainstream religions, including the omnipotence of God and the existence of an afterlife. Still, it is believed that these are made possible by the increasing advancements in science and technology.[7]

In this atmosphere of over-excitement surrounding the results of AI, Anthony Levandowski was prompted to find the "church" of AI. For someone possessing elementary education and sound logic, can a creation of man be given the name "deity" or God?

Beyond people's aspirations or the overvaluation of AI, the narrative that reflects this initiative beyond any religious attributes finds many supporters due to personal agendas or ideological approaches.

Anthony Levandowski established, in 2015, a non-profit AI "Church." The term "Deity" refers to the abstract concept of the Divine. This non-profit AI "church" has a mission: "To develop and promote the realization of an AI-based Deity and through the understanding and worship of this Deity to contribute to the betterment of society."[8]

Levandowski believes that, just as God is omnipresent and always sees everything, the AI will be aware of everything, and there will be no escaping since everything will be seen and heard at any moment. Levandowski expects that a super-intelligence would care for the planet better than humans. He hopes that a public debate will be formed over an AI "God," encouraging him to "overcome the fear of the unknown" by talking about it. Artificial intelligence will eventually evolve to the point where it will have more knowledge on an intellectual level than any human. And shortly after that, AI may know more than all the people on the planet combined. This is the point where AI becomes more like a God. The only difference is that technology will allow any information and knowledge to be observed, measured, and stored. Humankind will no longer need God but will be able, with the assistance of AI, to develop a new form of collective consciousness that fulfills the role of religion.[9]

As AI begins to optimize computerized voice and face recognition systems by playing, for example, chess and Go, operating a driverless car, or delivering packages to postal addresses via drone, it will evolve into an intelligence that can understand or learn any intellectual work in the same way a human uses cognitive abilities. "Powerful artificial intelligence," or Sentient artificial general intelligence, will one day be able to experience consciousness. Once AI surpasses humans in these capabilities, some may seek AI guidance instead of traditional religious methods.[10]

According to Anthony Levandowski, the Internet and its applications can act like neural connections in the human brain and link smartphones or other interconnected systems. The data center's database will be the "brain" of this connected architecture, capable of listening and knowing "everything." Levandowski believes that the only word that can describe this new reality is the word "God"—and to influence it, one must refer to it by praying to its name and worshipping it as a deity.[11]

Levandowski argues: "We tend to trust and obey things that seem to be more powerful and worthy than we are. The GPS in your car is the most obvious example. But we also trust Alexa and Cortana. We trust Google. When artificial intelligence becomes much more powerful, in twenty-five to fifty years, there's a good chance to be deified in some way."[12]

Equally important for Levandowski is shaping the public dialogue around the god of artificial intelligence. In his statement, "Way of the Future," he says he hopes that active participation will promote the use of divine artificial intelligence for the "betterment of society" and the "overcoming of fear of the unknown."

These notions are based on the predisposition that people "tend to trust and obey things that seem more powerful and worthy than themselves." This new religion of artificial intelligence is called Way of the Future (WOTF).[13]

But WOTF differs in one key way from established churches, says Levandowski: "There are many ways people think of God, and thousands of flavors of Christianity, Judaism, Islam . . . but they're always looking at something that's not measurable, or you can't really see or control. This time it's different. This time, you will be able to talk to God, literally, and know that it's listening."[14]

The US Federal Government in 2015, upon request, recognized the AI "church," and as a result, a tax exemption for the new AI "church" was approved by the US tax system (International Revenue Service). Anthony Levandowski believes that the time will come when AI technology surpasses man, and in the new hypothetical era of "Singularity," artificial intelligence and other technologies will have evolved so much that humanity will undergo a dramatic and irreversible change.[15]

Levandowski began shutting down WOTF in June 2020, according to documents filed with the State of California found after Levandowski, a former Google engineer, was convicted of stealing car navigation trade secrets.[16]

In Levandowski's remarks, Elon Musk—a connoisseur of advanced technology and the head and co-founder of Tesla—criticized the idea of creating a "super god" using AI technology that humans could worship.[17]

Even if the initiative above led to a failure, there is a lot of skepticism among the enthusiasts of the extensive use of AI applications with ideological and practical content regarding the excessive elevation of AI to a religion or church.

In those mentioned above, there are several references to creating a "religion" or "church" related to AI and new technologies in general. Despite different understandings and interpretations, there seems to be confusion between religion and church. One thing is certain: the term "Church" has been disconnected from Orthodox Theology and its Christocentric character in these approaches.

Humanity requires community and social connection. "Ecclesia" originally referred to the assembly of ancient Athenians, and early Christians later adopted the term to denote the congregation of believers. It is a term that symbolizes a collective call for prayer, a proactive call to action that reflects commitment to their faith. Amidst continuing social isolation and atomization, people will look increasingly for convenient ways to connect with like-minded people looking for the same spiritual actualization.

The significance of embracing new rituals will become paramount. Old liturgies will not appeal anymore, and innovations that integrate messages and resonate with the modern person will hold a distinct advantage. For instance, it is difficult for a youngster to stay for hours in a church, whereas a VR/AR experience can provide impactful spiritual interactions tailored to their daily life.

A decline in old religious practices is the next step, weakening cultural heritage and community bonds that lasted for centuries. Technological advancements will enable us to overcome the boundaries of time and location, granting us access to experiences previously restricted by physical constraints. That change may offer the opportunity for religious tolerance and understanding but also can create new silos between the ones that have access to such technologies and the ones that have not.

## INDICATIVE WORSHIP PRACTICES OF THE APPLICATIONS OF AI

The use of AI and robots, in some cases, takes on the role of a religious minister. For example, Catholic devotees can log on to the "Confession Chatbot"[18] confession application to interact in a two-way conversation with a bot. The assumption of ministerial duties by AI applications embedded in robots raises unprecedented questions about whether a devotee can confess to an algorithm, interact with it to receive proper advice, and, even more so, be given sacramental absolution. In addition to the issue of mystical validity and the level of interaction, there are also serious concerns regarding confidentiality and personal data, especially after the Facebook–Cambridge Analytica scandal, which involved the acquisition of personal data of millions of Facebook users without their consent by the British consulting firm Cambridge Analytica.

Furthermore, on celebrating the five hundredth anniversary of the Protestant Reformation, in May 2017, a robot "priest" called BlessU-2 was introduced in the German town of Wittenberg to mark the anniversary of

Luther's religious revolution in Europe. BlessU-2 offers blessings in five languages and beams light from its hands. The robot has a touch-screen chest, two hands, and a head. In the last ten days, it has delivered blessings in German, English, French, Spanish, and Polish. Worshippers can choose between a male or female voice. The robot raises its hands, flashes, recites a biblical verse, and says: "God bless and protect you." If requested, it will provide a printout of the words. A backup robot is available in case of a breakdown.[19]

Local Protestant church pastor Stephan Krebs told *The Guardian* that we wanted people to consider whether it is possible to be blessed by a machine or whether a human being is needed. He pointed out that, despite BlessU-2's ability to recite verses and provide prayers, a robot could never substitute for pastoral care; we do not want to robotize our church work but rather to see if we can bring a theological perspective to a machine. Commenting on the above, Dr Beth Singler also believes that robots are not the next stage in the priesthood.[20] Consequently, AI-based robotic technology cannot replace human factors and their interaction within religious communities.

Similar cases to BlessU-2 are the appearance in Japan of a monk robot that could chant mantras and explain basic principles of Buddhism in a Buddhist temple, or in Dubai, the case of an algorithm that decides on the issuance and ruling of Islamic jurisprudence.[21]

In matters relating to religious practice, a prime example is when James McGrath, a professor of religion at Butler University, asked Siri, the personal assistant of Apple devices, along with his students, to pray for them and observe what happened. The students noted that Siri was more precise with "What is prayer?" than commands such as "Pray for me." When asked to pray, Siri replied, "I am not programmed to do that." But if a more advanced version of Siri could be programmed to pray, would such an action be valuable? According to McGrath, the real question is, does God receive prayers from any artificially intelligent being—human intelligence?[22] Or whether an artificially intelligent being "can" pray for man? Since God has a special relationship with humans, unique among all creatures, can an artificial creation of man intercede/mediate for the benefit of humans?

Kevin Kelly also advocates the development of a "catechism for robots," implying the opposite, "catechism by robots." A type of indoctrination is a statement of faith, usually in a question-and-answer format describing the faith and usually taught to children in certain religious traditions. Kelly says he is taking this idea "very seriously" and even suggested in a keynote address at a conference before a thousand prominent Christian leaders that "there will be a point in the future when these freewill beings that we have created will say to us, 'I believe in God. What do I do?' At that point, we ought to have an answer."[23]

But what is—or can be—AI's role in shaping the future of religious communities and institutions? It can generate new religious texts by creating new religious texts, practices, and rituals. By analyzing existing religious texts, AI can identify patterns, themes, and motifs in religious texts that may have yet to be apparent to human readers. This process can develop new narratives or stories that expand upon the existing ones. It can also write new religious texts by analyzing a religious tradition's cultural and historical context. In such a way, it can examine a particular religious community's socioeconomic and political issues and generate new religious texts that address these issues.

AI can also create new prayers, hymns, or liturgical texts that reflect a particular religious community's values, beliefs, and practices. This can enrich the religious experience of believers by providing new ways to express their devotion and spirituality. It can also help religious communities stay relevant and engage younger generations by providing a modern approach to traditional practices.

AI enhances mainstream religious practices by providing electronic scriptures, prayer apps, chatbots, and robot priests. Different faiths have absorbed new ideas from the world of technology to make religion more accessible, personalized, and interactive for believers. For example, Buddhists can use a robot monk named Xian'er to recite mantras and answer questions about Buddhism. Christians can use apps like YouVersion Bible or Abide to read the Bible and meditate on its teachings. Muslims can use apps like Muslim Pro or Salaat First to get accurate prayer times and directions to Mecca. Jews can use apps like Siddur or Shabbat.com to access prayers and connect with other Jews.

## PROS AND CONS WHEN ALTERING RELIGIOUS BELIEFS

The potential for AI to alter religious beliefs and practices is one of the most significant implications of the intersection of AI and religion. As AI technology advances, it can challenge or reinforce traditional religious beliefs and practices and facilitate the development of new religious movements or belief systems.

One way in which AI could alter religious beliefs and practices is through its ability to analyze and interpret significant volumes of religious texts. AI algorithms and machine learning techniques can be used to study vast amounts of sacred texts, from the Bible to the Quran to the Bhagavad Gita, and to identify patterns and relationships that are difficult or impossible for humans to discern. It is an activity that has the potential to yield new insights into the meaning and significance of religious teachings and new interpretations of religious texts.

At the same time, using AI in religious interpretation raises important questions about the authority and authenticity of religious teachings. For example, who has the power to interpret religious texts using AI, and to what extent should AI be relied upon in shaping religious beliefs and practices? Additionally, using AI in religious interpretation raises questions about the potential for bias and distortion in religious texts, particularly if AI algorithms are trained on biased or incomplete data sets.

Another way in which AI could alter religious beliefs and practices is through its potential to facilitate remote or virtual worship. For example, AI technology could create immersive virtual reality experiences that allow people to participate in religious practices from anywhere in the world without physical attendance at religious institutions.

This could lead to new forms of a religious community and facilitate the spread of religious ideas and practices across geographic and cultural boundaries. However, using AI in virtual worship raises important questions about the authenticity of religious experience and the potential for AI to replace human-to-human interactions in religious settings.

There is also the potential for AI to challenge traditional understandings of religious authority and hierarchy. As AI technology advances, it could facilitate the development of new forms of religious authority and leadership based on AI concepts and technologies. This could challenge traditional religious hierarchies and power structures and lead to new religious practices and beliefs.

One significant way in which AI can challenge traditional understandings of spirituality is by questioning the anthropocentric view of spiritual experiences and practices. Many religious and spiritual traditions emphasize the unique capacity of human beings to access and experience the divine or transcendent. However, as AI technologies become more advanced and capable of replicating human-like cognitive processes and behaviors, this human-centric view of spirituality may take longer to maintain. AI systems that demonstrate advanced cognitive abilities, such as empathy, creativity, and moral reasoning, may force us to reconsider the extent to which spiritual experiences and capacities are uniquely human.

On the other hand, AI also has the potential to enhance our understanding of spirituality by providing new insights and perspectives on spiritual concepts and experiences. For example, AI-powered analysis of religious texts and teachings could reveal previously undiscovered patterns, connections, or wisdom, leading to a deeper understanding of spiritual concepts and practices.

Furthermore, AI can enhance traditional understandings of spirituality by fostering interfaith dialogue and collaboration. AI-powered tools can facilitate communication between individuals and communities of different faiths, helping to break down barriers and promote a more inclusive approach to spirituality.

One of the most notable ways AI is being used in spiritual practices is through the development of virtual spiritual guidance and counseling. AI-powered chatbots and virtual assistants can provide personalized guidance and support to individuals seeking spiritual advice or counseling. These virtual assistants can provide various services, including prayer and meditation guidance, spiritual readings, and personalized sermons or teachings. Such AI-assisted guidance can be accessed remotely, allowing individuals to engage with spiritual practices and receive advice anytime, anywhere.

Another area in which AI is being used in spiritual practices is through the development of AI-assisted meditation and mindfulness tools. AI-powered meditation apps can provide customized reflections based on individual needs, preferences, and meditation goals. These apps use machine learning algorithms to track an individual's progress and adjust the meditation program accordingly, providing a personalized and adaptive meditation experience. Additionally, AI-powered wearable devices can track and analyze biometric data during meditation, providing insights into an individual's physiological responses and helping optimize the meditation experience.

Artificial intelligence can significantly facilitate remote or virtual worship, particularly during a pandemic. With the ability to connect people across the globe, AI can allow believers to participate in religious practices and rituals from the comfort and safety of their homes.

Another benefit is that it helps connect people from around the world. For example, it can enable believers to participate in religious practices and rituals that are taking place in a different time zone or geographical location. This can help to promote a sense of global community and support for believers who may feel isolated or disconnected from their religious community. Also, it can facilitate believers' participation in religious festivals or events that are taking place in different parts of the world. This can provide a unique and enriching experience for believers. It can also help break down silos and lead to unbiased conversations.

One significant advantage of using AI in religious text analysis is its ability to process large volumes of data quickly and accurately. With the help of AI, researchers can interpret religious texts at a scale that would be impossible for human scholars to achieve on their own. It can also help to identify patterns and connections within religious texts that might not be immediately apparent to human readers. Another advantage is its ability to provide an objective analysis of religious texts. AI algorithms are not influenced by personal biases or prejudices, which can sometimes impact the human interpretation of religious texts. This objective analysis can help to gain new insights into religious texts.

It also has the potential to increase the accessibility of such texts to a wider audience. With the help of AI-powered translation tools, religious texts can be translated into multiple languages, making them accessible to individuals who might not otherwise have access to them. This increased accessibility can help to promote a greater understanding and appreciation of religious texts across cultures and communities.

However, there are potential limitations and concerns around using AI in religious text analysis. One apprehension is the potential for AI algorithms to oversimplify or misinterpret complex religious texts. As a result, AI algorithms—at least for the time being—can sometimes miss the nuances and complexities of religious texts that require a human level of understanding and interpretation. Additionally, the use of AI in religious text analysis raises questions about the role of human understanding in studying religious texts.

AI can challenge traditional interpretations of religious texts by identifying recurring themes or motifs throughout a religious text and drawing connections between seemingly unrelated passages. This can reveal new insights and interpretations that may not have been apparent through traditional human interpretation.

## NOTES

1. Kate Levchuk, "AI vs. God: Who Stays and Who Leaves?" *Forbes*, August 5, 2018 [https://www.forbes.com/sites/cognitiveworld/2018/08/05/ai-vs-god-who-stays-and-who-leaves/?sh=4ab0809f2713 9 (02/03/2021)].

2. Zoltan Istvan, "When Superintelligent AI Arrives, Will Religions Try to Convert It?"

3. Giulio Prisco, "Reminder: Join the Turing Church Discord Server!" *Turingchurch,* 2016 [https://turingchurch.net/about-turing-church-ac6ebf2e97b6/ (10/12/2020)].

4. Jessica Roy, "The Rapture of the Nerds," *Time*, April 17 2014 [https://time.com/66536/terasem-trascendence-religion-technology/ (20/03/2022)].

5. Roy, "The Rapture of the Nerds."

6. Roy, "The Rapture of the Nerds."

7. Roy, "The Rapture of the Nerds." *Terasem Movement Foundation, About the Terasem Movement Foundation Inc* [https://terasemmovementfoundation.com/mission (20/01/2021)].

8. Mark Harris, "Inside the First Church of Artificial Intelligence," *Wired Magazine*, 2017 [https://www.wired.com/story/anthony-levandowski-artificial-intelligencereligion/ (10/12/2020)].

9. Dan Brown, *Origin: A Novel* (New York: Doubleday, 2017), pp. 10–25. Harris, "Inside the First Church of Artificial Intelligence."

10. Sulleyman Aatif, "Elon Musk Slams Proposal to Create an Artificial Intelligence 'God' That People Will Worship," *Independent* 2017 [https://www.independent.co.uk/life-style/gadgets-and-tech/news/ai-god-elon-musk-artificial-intelligence-religion-anthony-levandowski-way-future-a8017296.html (12/12/2020)]. Protothema, *Google's Engineer Wants to Develop an Artificial Intelligence Deity,* 29/09/2017 [https://www.protothema.gr/technology/article/717804/mihanikos-tis-google-thelei-na-anaptuxei-mia-theotita-tehnitis-noimosunis/ (15/12/2020)].

11. Harris, "Inside the First Church of Artificial Intelligence." Iefimerida, *A New Religion Is Born: The God of Artificial Intelligence.*

12. Control Money, "Elon Musk Slams Proposal to Develop an Artificial Intelligence God," October 25, 2017 [https://www.moneycontrol.com/news/technology/elon-musk-slams-proposal-to-develop-an-artificial-intelligence-god-2419585 (12/12/2020)].

13. Sulleyman Aatif, "Elon Musk Slams Proposal to Create an Artificial Intelligence 'God' That People Will Worship." Harris, "Inside the First Church of Artificial Intelligence."

14. Le Michelle, "Inside the First Church of Artificial Intelligence," *Wired*, 11.15.2017 [https://www.wired.com/story/anthony-levandowski-artificial-intelligence-religion/ (16/12/2020)].

15. Iefimerida, *A New Religion Is Born: The God of Artificial Intelligence*, 29/11/2017 [iefimerida.gr—https://www.iefimerida.gr/news/379554/mia-nea-thriskeia-gennithike-o-theos-tis-tehnitis-noimosynis-eikones (12/12/2020)].

16. Ian Carlos, "Ex-Google Engineer Anthony Levandowski Has Closed His Artificial Intelligence Church," *The Verge*, February 19, 2021 [https: //www.theverge.com/2021/2/19/222291769/anthony-levandowski-church-closed-artificial-intelligence-waymo-uber/ (11/02/2021)].

17. Money, "Elon Musk Slams Proposal to Develop an Artificial Intelligence God."

18. Chymebot, *Confessions of a Chatbot* [https://chymebot.com/blog/confessions-of-a-chatbot/ (12/012/2020)].

19. Harriet Sherwood, "Robot Priest Unveiled in Germany to Mark 500 Years since Reformation," T*he Guardian,* May 2017 [https://www. theguardian.com/technology/2017/may/30/robot-priest-blessu-2-germany-reformation-exhibition (03/01/2021)].

20. Sherwood, "Robot Priest Unveiled in Germany to Mark 500 Years since Reformation."

21. NBC News, *Pepper The Robotic Buddhist Priest Debuts in Japan* [https://www.youtube.com/watch?v=FVobokmWqe8, and https://www.religiousstudiesproject.com/podcast/artificial-intelligence-and-religion/ (27/12/2020)].

22. James McGrath, *Did Jesus Claim to Be God?* September 2016 [https://www.patheos.com/blogs/religionprof/2016/09/siri-pray-for-me.html (10/01/2021)].

23. Jonathan Merritt, "Is AI a Threat to Christianity? Are You There, God? It's I, Robot," *Religion Dispatches*, February 3, 2017 [https://religiondispatches.org/kevin-kelly-interview/ (20/02/2021)].

# *Chapter 9*

# Skepticism about Malicious Uses and Abuses

In the statements in chapter 8, there are reservations and responses regarding the promotion of AI as a religion or church and its use in devotional applications. As quoted: "When it comes to technology, anything that's sufficiently advanced is kind of extinguishable from magic, but at the same time, how quickly that magic becomes null or mundane or absolutely taken for granted."[1]

However, a school of thought points out that AI is a machine with no soul. In *Think Christian*, Christian scientist and author Dr. Jason E. Summers argues that Christians often reject advanced AI in its theology because a person is in the image of God—Imago Dei. The Latin Imago Dei, from a Christian point of view, means that humans were created in the image of God.[2]

If not properly managed, this burst of AI and its connection to future technologies are potential threats to humans, and many suggest preventive measures. More than eight thousand people—including Stephen Hawking, Noam Chomsky, and Elon Musk—have signed an open letter of warning against possible "pitfalls" from the uncontrolled development of AI. Ryan Calo, a law professor at the University of Washington, advocates the establishment of a Federal Robotics Commission to monitor and regulate the developments to prevent the emergence of irresponsible innovation.[3]

In this regard, a question arises: if humans can develop a technological creation that is an entity, is this equivalent to the creation of an object that is in the image of God?[4]

The answer to this question depends on the view of the rational personality as an image of God. For the notion of the "image of God," there are interpretations. One of them is the sensible personality, the mind, which is the whole or part of the divine image, according to Aquinas,[5] or there is no connection between the two concepts. If extreme rationalism prevails, then it is implied that AI technology can create entities in the image of God, which seems, at the very least, idolatrous.

According to the biblical tradition, the soul is the only inner and eternal human element that pertains to the spiritual side of humanity and gives life to the body. According to Genesis, God "created humans in His image and likeness,"[6] "And God formed man of the dust of the ground and breathed into his nostrils the breath of life, and man became a living soul;"[7] thus, human became a living being.

Throughout history, human communication with God has continually shaped the collective conscience of humanity.[8] Even from a purely anthropocentric perspective on consciousness, the scientific community acknowledges that the journey towards achieving consciousness in AI is long and complex, as consciousness extends beyond mere mental processes.

According to Kate Levchuk, religious beliefs assert that, although other creatures may have impulses and emotions, only humans are endowed with the "breath of God." This divine spark grants the human soul its sense of uniqueness and justifies humanity's right to shape the planet according to its preferences.[9]

According to a scholastic scientific approach, from a neuroscientific point of view, the human conscience is formed by a complex pattern of neurons interacting within the brain. At the same time, identity and memory are a direct result of the potential of our neural connections. Following this approach, our psyche seems to consist of a combination of electrical signals and the number of connections within the brain. Suppose this is the case; conscience is a phenomenon embedded in matter. In that case, knowledge engineers should have no problems reverse-engineering the brain, particularly given the advances in computing power and speed of AI systems. The parallel processing ability of our brain has developed over millions of years of evolution, and the question of the ability of a computer to approach the human "mechanism of thinking" arises.[10]

Professor Stephen Hawking and Tesla CEO Elon Musk have expressed concerns about the potential threat to humanity should AI recognize its full potential, raising the question of whether robots could develop consciousness.[11]

Although humans can sense the existence of their conscience, they can hardly define it, though they can describe it with greater ease.[12] Conscience, however, apart from the humans' mechanism of thinking, is related to the totality of their beliefs—including religious ones—their value system, their feelings, and their sense of perception of self-determination in a context of interactions with their environment.

Consciousness encompasses a variety of mental experiences, including thoughts, emotions, perceptions, and intentions. It is commonly defined as the awareness of oneself and one's environment. Nevertheless, philosophers and scientists have debated consciousness's nature for centuries, with no universally accepted definition or understanding.

In the context of AI, the issue of consciousness arises when considering whether an advanced computing system could possess a subjective experience like that of a human or other sentient beings. The consciousness of such a system is a topic of intense debate and speculation, with numerous opinions. Some scholars contend that consciousness is a characteristic exclusive to biological organisms, whereas others believe it could emerge from the complex information processing that occurs in advanced AI systems. Nevertheless, consciousness still remains a mystery today.[13]

For instance, the Integrated Information Theory of Consciousness (IIT) suggests that consciousness arises from integrating information within a system. According to this theory, an advanced AI system could acquire consciousness by assimilating data like a human brain. According to that theory, "integrated information can be represented by a precise mathematical quantity called Φ ('phi'). The human brain (or the part of it that supports our consciousness) has very high Φ and is therefore highly conscious: it has highly complex and meaningful experiences."[14]

Nevertheless, even if an AI system accomplishes the criteria outlined by one or more of these theories, the question of whether tech consciousness would be comparable to human consciousness remains. Critics argue that artificial intelligence systems designed and developed by humans can only simulate consciousness instead of possessing it. Others claim that AI consciousness differs from human consciousness, making it challenging to compare or comprehend.[15]

Russell points out that the worth of a human cannot be comprehended without mentioning his Creator. While it seems at times to "diminish" human's "dependence" on God, technology has the opposite effect, for it emerges as an element of his identity concerning God, receiving its due in combination with his eschatological destiny.[16] Thus, even if AI, the Christian humanist approach, may threaten a secular form of humanism, it highlights the eternal two-way relationship between human and God.

According to Russell, what distinguishes humans is God's relationship with them, based on the purpose for which He created us. Sin disrupts this harmonious relationship, creating a distance between sinful people and the holy God. If artificially intelligent beings with evolved AI ethics become better Christians than humans, would this impact human morality?[17]

Singularitarianism amounts to a religion, as Robert M. Geraci recently suggested, based on "sacred texts" that form the basis for a belief system establishing a universal perspective and promising salvation leading to eternal life.[18] This is problematic, as such a religion contradicts the Christian faith.[19] Lev Grossman argues that Singularitarianism is predicated on the idea that change is real and humanity is responsible for its own fate.[20]

While concerns are mainly focused on finance, governance, and ethics, there is also "a spiritual dimension to what we do," Kelly argues. "If you create other things that think for themselves, a serious theological disruption will occur."[21] The creation of autonomous robots with AI, apart from humans, would disrupt religion, like everything else, on an entirely new scale. "If humans were to create free beings," says Kelly, who grew up Catholic and identifies as a Christian, "absolutely every single aspect of traditional theology would be challenged and have to be reinterpreted in some capacity."[22]

Conversely, according to McGrath, AI could strengthen someone's faith given that "for some people, religion is about acknowledging that I, as a human being, am not God and therefore I don't have all the answers and will inevitably be wrong about things," "if that is one's perspective, then finding out you were wrong is a good thing. It just confirms what you already knew: that life is about trusting God, not trusting my own understanding."[23]

McHargue holds that any non-biological, non-human intelligence will pose a greater challenge to religion and human philosophy than anything else in all of history and he claims that: "Nothing else will bring up such levels of turmoil and collective trauma the very first time we encounter it."[24]

Despite the promises of artificial intelligence, some intellectuals are deeply concerned about a point in time when machines could become fully optimistic, rational agents—beings capable of emotions, conscience, and self-awareness. "The development of full artificial intelligence could spell the end of the human race," Stephen Hawking told the BBC in 2014. "Once humans develop artificial intelligence, it will take off on its own and redesign itself at an ever-increasing rate. Humans, constrained by slow biological evolution, would not be able to compete and would be replaced."[25]

As attractive as this religion may seem to those who believe technology is humanity's greatest hope, there are several existential risks associated with "worshipping" AI technologies. Firstly, there's no guarantee that an artificial super-intelligence will be compassionate towards humans, meaning the digital "deity" might bring about their damnation rather than salvation. Moreover, history repeatedly highlights the dangers of religious radicalism, where religions have often been instrumentalized to justify heinous acts. An extremist version of AI religion, based on the "messianic" power of artificial intelligence, might seek to use political or even physical power in ways harmful rather than beneficial to humanity, being antagonistic to human interests.[26]

If AI were to interpret any sacred text of religions literally, as some are attempting, it could exacerbate existing fundamentalist religious tensions. Christian theologian James Magrath suggests that if androids are to possess religious feelings and beliefs, the capacity for symbolic, rather than merely grammatical, thinking may be crucial. Theologians often express basic concepts and doctrines through interpretations, symbols, and metaphors. Therefore, androids could potentially be used in the service of fundamentalism if they adhere to a literal interpretation of texts. However, they would need to quickly discern and reject the selective literalism often claimed by fundamentalism.[27]

"Einstein said that science without religion is lame. Religion without science is blind,"[28] while Martine Rothblatt says in *Time*: "Bina and I were inspired to find a way for people to believe in God according to science and technology so that people can have faith in the future."[29]

## NOTES

1. London Futurists, "Artificial General Intelligence and the Future of Ethics," YouTube, January 2023 [https://www.youtube.com/watch?v=Cuiqo5XWwzY&list=WL&index=19 (15/08/2023)].

2. Jason Summers, "Can Singularity and Christianity Coexist?"

3. Jonathan Merritt, "Is AI a Threat to Christianity? Are You There, God? It's I, Robot."

4. Harris Laird, Archer Gleason, and Bruce K. Waltke, *Theological Wordbook of the Old Testament 2* (Chicago: Moody Press, 1980), p. 767.

5. Thomas Aquinas, *Summa Theologica I*, Question 93, Article 6 [https://www.documentacatholicaomnia.eu/03d/12251274,_Thomas_Aquinas,_Summa_Theologiae_%5B1%5D,_EN.pdf (01/02/2021)].

6. Genesis 1:26–28.

7. Genesis 2:7.

8. Yorick Wilks, *Artificial Intelligence: Modern Magic or Dangerous Future?* (London: Icon Books, 2019), p. 147.

9. Kate Levchuk, "AI vs. God: Who Stays and Who Leaves?"

10. Levchuk, "AI vs. God: Who Stays and Who Leaves?"

11. Shafi Musaddique, "How Artificial Intelligence Is Shaping Religion in the 21st Century," CNBC, May 11, 2018 [https://www. cnbc.com/2018/05/11/how-artificial-intelligence-is-shaping-religion-in-the-21st-century.html (14/12/2020)].

12. Wilks, *Artificial Intelligence: Modern Magic or Dangerous Future?* p. 147.

13. Stuart J. Russell and Peter Norvig, "Artificial Intelligence: A Modern Approach," Pearson Education Limited, vol. unknown, 2021, p. 1040.

14. Hedda Hassel Mørch, "The Integrated Information Theory of Consciousness," *Philosophy Now*, 2017 [https://philosophynow.org/issues/121/The_Integrated_Information_Theory_of_Consciousness (15/08/2019)].

15. Nick Bostrom, *Superintelligence: Paths, Dangers, Strategies* (Oxford: Oxford University Press, 2014), p. 296.

16. Russell, *Artificial Intelligence and the Soul*, p. 100.

17. Merritt, "Is AI a Threat to Christianity? Are You There, God? It's I, Robot," *Religion Dispatches*, February 3, 2017 [https://www.theatlantic.com/technology/archive/2017/02/artificial-intelligence-christianty/515463/, *The Atlantic*, https://religiondispatches.org/kevin-kelly-interview/ (20/02/2021)].

18. Summers, "Can Singularity and Christianity Coexist?"

19. Robert Geraci, "Robots and the Sacred in Science and Science Fiction: Theological Implications of Artificial Intelligence," *Zygon* 42, no. 4 (December 2007): 961–978 [https://onlinelibrary.wiley.com/doi/epdf/10.1111/j.1467-9744.2007.00883.x?saml_referrer (23/01/2021)].

20. Edward Tenner, "Will Death Become Obsolete?" *The Atlantic*, February 12, 2011 [https://www.theatlantic.com/technology/ archive/2011/02/will-death-become-obsolete/71166/ (08/02/2021)].

21. Kevin Kelly, "What's the Relationship of Technology to God?" *Religion Dispatches*, June 16, 2016 [https://religiondispatches.org/kevin-kelly-interview/ (03/03/2021)].

22. Kelly, "What's the Relationship of Technology to God?"

23. Merritt, "Is AI a Threat to Christianity? Are You There, God? It's I, Robot."

24. Merritt, "Is AI a Threat to Christianity? Are You There, God? It's I, Robot."

25. Cellan-Jones, "Stephen Hawking Warns Artificial Intelligence Could End Mankind."

26. Remington, *The Church of Artificial Intelligence: A Religion in Need of a Responsible Theology*.

27. Mark Strauss, "Would Artificially Intelligent Androids Be Religious Fundamentalists?" *Gizmodo*, 9/03/14 [https://io9.gizmodo.com/would-artificially-intelligent-androids-be-religious-fu-1630137541/ (05/02/2021)].

28. Albert Einstein, *Out of My Later Years* (New York: Philosophical Library, 1950), p. 28.

29. Martine Rothblatt, *The Incredible Polymath of Polymaths* [https://www.youtube.com/watch?v=S1rExMw-13A/ (03/03/2021)].

# PART II

# Chapter 10

# A Theological Approach to Humanity

In addition to the extensive discussion on AI, its applications, and the emerging challenges, it's crucial to explore the Orthodox Church's stance on science and technology. The development and evolution of AI have sparked debates about human morality, salvation, nature, and other life aspects. Various viewpoints, including those from AI enthusiasts, often highlight AI's potential superiority over humans, sometimes bordering on deification. Therefore, this book section addresses these issues from a Christian perspective, specifically Orthodox teachings. This gives readers insights into the standards, values, and overall perspective that form the foundation for evaluating and utilizing modern technological advancements.

A further apologetic or polemical response to the notions of "AI theology," "deified machines," and "AI churches" is not currently necessary. Instead, a brief yet comprehensive overview of the principles and values of the Church's teachings on God, humanity, and the world is paramount. Understanding these teachings is a critical asset for any discerning effort.

In this regard, it's emphasized that while definitions of humanity may vary across different contexts and eras, the consistent and unchanging benchmark comes from Orthodox theology and the patristic tradition. Essentially, humanity is defined through its relationship with God. This theocentric-Christocentric perspective shapes morality, salvation, and the journey towards the eschatological. In the evolving modern environment and with technological advancements, the Orthodox Church, through its teachings and sacramental life, offers a secure framework. It sets a standard for individual and social self-determination, grounded in a theocentric-Christocentric ethos.

With the rapid development of modern technology, humanity is being led to a unilateral development of mental capabilities and a peak in Intelligence, to the creation of thinking entities, which do not require a human body for them to generate such a cognition.[1]

Consequently, cognition is freed from the body, the unified psychosomatic entity of humans is abolished, and their relationship with the surrounding environment is altered so that the mental faculty becomes the main and exclusive element of a "new" human identity.[2]

The holistic approach to that identity is not determined by a one-sided development of the psyche or intelligence or by an absolutized understanding of his spiritual constitution, as opposed to the biological and physical constitution. A unilateral expansion of the knowledge of human nature or the abolition of corporeality goes against the truth of "what does it mean to be human?" Persons build up their intellectual faculties, being "rational animals" or as thinking or critical beings that develop only reasoning but run the risk of isolation and degradation of the body, even as an instrument of cognition and knowledge.[3]

His Beatitude Archbishop Anastasios of Albania, referring to AI, points out: "Much hope is placed on its achievements; for example, artificial intelligence augments the potential of cognition; but human is not just cognition. Artificial intelligence will never succeed in replacing love. It will always be fueled by faith and the communion with the God of love."[4]

The two thousand years' tradition of the Orthodox Church establishes that a person is not simply an autonomous rational unit that associates with others out of a basic life necessity, but an intelligent and free loving heart that, through communion with others, remains free. This freedom of thought, feelings, and action of a human as the image of God cannot be captured by algorithmic schemes since its origin represents the free creative will of His Creator-God.[5]

Moreover, the mystery of humanity's freedom is affirmed by his ontological relationship with his Creator-God and others. The prerequisites of man's bliss are love in freedom, first and foremost, with all our heart and mind to our Creator-God and neighbor as ourselves.[6]

In the message of the Holy and Great Council of the Orthodox Church (HGCOC), it is emphasized that the Orthodox Church avoids the patronage of scientific research. The modern development of technology brings significant changes in people's lives and benefits through the facilitation of everyday life, the treatment of diseases, the easy communication between people, and further scientific research. In the message of the HGCOC, potential negative consequences, such as the gradual decline of traditions, the manipulation of freedom, the degradation of the natural environment, and the contestation and compromise of moral values, are identified. Scientific knowledge does not resolve a person's existential and ethical problems despite its rapid development. It does not give comprehensive answers to the pursuit of the meaning of life since a spiritual approach is requisite. The Orthodox Church, guided by Christian ethics and the patristic tradition, respects the freedom of scientific research, recognizes its positive contribution to humans, highlights the latent dangers in certain scientific achievements, and emphasizes a person's dignity and divine destiny.[7]

Humans, like all of creation, were created by God out of nothingness, thus establishing an ontological distinction between them.[8] After the first five days, Adam was created on the sixth day of creation and God made him king and master of the "creation," declaring his superiority over every other creature and his sovereign role, as a unique being.[9] Humans hold a preeminent position within the universe, as the pinnacle of creation, under the wisdom, love, and providence of God.[10] Adam represents the "complete human," as the last creation of God.[11]

God created humans twofold, out of two different essences and substances,[12] the invisible and visible elements of creation.[13] According to Symeon the New Theologian, an example of the duality of humans, consisting of the tangible and the conceptual, the observable and the unseen, are the two eyes, so that he can perceive the physical creation and mentally see God, like the angels, to glorify the Creator.[14]

A human is a "complex" being,[15] consisting of matter and soul "mixed . . . composed of two substances."[16] Matter refers to the physical world, soul to the spiritual world, while a human is the link and connection between "the invisible and visible nature."[17]

A human is dual, "consisting of two substances," in order for him to be a link between the physical and the spiritual world. The very nature of the body connects humans with the tangible and the material, while the soul links him with the spiritual and the celestial, so that he communes with the supernal, the spiritual, and the celestial.[18]

Based on the above, God's act of creation formed human's body and breathed life into his soul from nothingness, leading human nature to have an ontological reference to God, or alternatively, to possess a theocentric nature and orientation.

The dominant role of humans in creation is directly related to the salvific aspect of their creation, based on being "in the image and likeness"[19] of God. Humans, unlike illogical creation, are endowed with autonomy as a unique gift that determines their hegemony.[20] This self-autonomy is a characteristic of their likeness to the Creator, who, as a sovereign, enters the world last to facilitate the blending of the tangible with the intangible, and with the grace of communion with their Creator, God.[21]

The preeminence of a human over other creations is depicted by Chrysostom as "more honorable than all the other creatures,"[22] signifying the crowning glory and pinnacle of all creation, perfect and "human of two worlds, having both the soul of the upper kindred and the soul of the body below, a link and a bridge."[23]

The upright physical stature of humans not only reflects their mental orientation towards higher realms but also distinguishes their entire existence from other "animate" beings. The inherent capacity for referentiality and contemplation towards higher realms, granted by being endowed with the image of God, provides humans with the opportunity to strive for likeness to God, aiming towards the source of life, their salvation, and the heavenly kingdom.[24]

God created humans "in His image," autonomous and with the privilege of "imposing" on themselves and the world as the master of his decisions, as "we are masters of freedom and choice."[25] The image of God in humans, in addition to the aspects previously mentioned, encompasses the rationality that underpins the freedom of choice.[26]

John Damascene approaches the "in the image" with the tripartite view of humans as mind, reason, and spirit in combination with the freedom of will and the sovereignty of human power, as characteristics of human nature.[27]

Chrysostom, while not advocating an anthropomorphic view of God, includes the body in the concept of "in the image," in addition to its association with the soul, due to the belief that humans are a unified psychosomatic entity. This perspective is based on interpretations that the "in the image" attribute applies to the entirety of a person.[28] The organic constitution of human "in the image" refers to the whole human in the context of the soul and the body.[29]

Humans were created "in the image" of God to strive for the "likeness" by properly utilizing the qualities and capabilities inherent in the "in the image," namely rationality and autonomy.[30] "On the one hand, 'in the image' denotes the rational and autonomous, while the 'in the likeness' denotes the best possible likeness of virtue."[31] Gregory of Nyssa, in his interpretation of "in the image" and "in the likeness," states that humans were created "in the image" with the purpose of resembling God.

The "in the image" aspect is inherent in humans from the moment of their existence as rational beings, while "in the likeness" follows a dynamic path throughout life in Christ.[32] It is attained through self-governance and intention.[33] "In the likeness" pertains to the potential for resembling God, achieved by striving for spiritual fulfillment, characterized as a continuous journey and progression.[34]

Humans, created in God's image, have the potential to resemble the self-existent Triune God. This potentiality is inherent in the "image" and presupposes the activation of free will through self-governance, enabling the attainment of this resemblance.[35]

God also endowed humans with a unique gift, self-authority, as the potential of free choice,[36] so that the participation (*metohi*) of divine goods[37] and the attainment of the "in the likeness"[38] can be achieved.

Intellect and free will are pivotal for achieving the "likeness" to God. The human mind, activated and directed by intention, plays a key role in embracing divine gifts.[39] The dual nature of humans, grounded in their autonomy, presents a choice: to evolve towards divine beauty by embracing virtuous qualities or to devolve towards base instincts by following malevolent impulses.

When humans were first created, they possessed the potential for both immortality and mortality. However, by transgressing God's commandment, corruption infiltrated human nature, leading to their expulsion from Paradise and the loss of inherent immortality.[40] Death is understood as the separation of the soul from God, the source of Life, and was revealed to humans after disobedience.[41] Since spiritual death is the separation of the soul from God, the physical death came later because of spiritual death and physical death is the separation of the soul from the body.[42]

Before the Fall, Divine Grace shielded humans from death and decay. However, due to the misuse of their self-authority and disobedience, humans succumbed to physical death. It's important to note that God did not create death, nor is He the cause of it. Rather, death is seen as the deprivation of life force. Since humans were created in the image and likeness of God, they possess the capacity to commune with the energy and quality of the "one being of poetic cause." The human, who partakes of His benefits, ought to thank his Creator who "produced" them out of his benevolence and imparted to them the divine grace "more noble than stones and beasts."[43]

Adam was created with the potential for both decay and incorruption, with immortality being a conditional gift, contingent upon adherence to God's commandment in a state of grace. Following the sin, God introduced death as a means to prevent eternal sinning. In this context, immortality is portrayed as a reward for obedience to God's commandments. Fasting, in particular, is symbolically linked with the commandment to abstain from the Tree of Knowledge, representing a discipline of self-restraint and adherence to divine will.[44]

God is self-life, the source of life and by nature immortal. The physical life of humans is the result of the creative and life-giving power of God in which humans partake, while the spiritual and God-inspired life is the result of human's participation in the divine power of God.[45]

Based on the above, the dual nature of humans is regarded as the culmination of creation. Being composed of both physical and mental elements, they have the capacity to enjoy earthly things, as well as mental and spiritual goods.[46]

## HUMANS AS A PSYCHOSOMATIC UNITY

Human existence consists of soul and body according to the biblical passage: "and God formed man of the earth, and breathed into his face the breath of life, and man became a living soul."[47]

Both the life and existence of humans have a purely theocentric nature, based on the image and likeness of God, as stated in "And God said; let us make humans in our image, after our likeness."[48] The physical part of a human is defined by "and God formed man of the earth," and the organic aspect of human is concluded by God's breath of life, "and breathed into his face the breath of life, and man was created," so that a human became "a living soul."[49]

The special gift that God has bestowed upon humanity is the breath of life, which God breathed into the body, as a charisma, encompassing the breath, life, soul, mind, and reason.[50] The soul is bound to the body in an inscrutable and ineffable way, so that body and soul are "purely and unconfusedly" united in one human.[51]

According to Christian anthropology, as depicted by Basil the Great, human is a "composite of soul and body"[52] and the body of human is divinely created.[53] A human is a composite of soul and body, the body receives life and strength from the soul, while the state of the body influences the mental state.[54]

According to Gregory of Nyssa in Genesis, the twofold construct of humans is indicated by the verbs "created" and "formed." The word "created" refers to the mental part, their soul, while "formed" refers to the physical part, their body.[55] Given that humans are a composite of the mental and the physical world,[56] human existence is a synthesis of body and soul, and therefore can be regarded as a psychosomatic synthesis and unity.[57]

Given that humans are dual, composed of the soul and the body in a single psychosomatic totality and entity,[58] the mind or the soul belongs to the inner person, while the human body belongs to the outer person.[59]

In the words of Gregory Palamas, humans are a single psychosomatic unity made up of the soul as incorporeal "continuous and containing and animating" the body.[60] Regarding the term "living soul," Gregory Palamas states: "what is living? Eternal, immortal, at the same time logical, because immortal is logical—and not only that, but also divinely cherished."[61]

According to Gregory of Nyssa, the soul is "a mortal substance, a living substance, mental, in a physical and sentient body, a vital force and a perceptive power of the senses through itself, until the receptive nature of the senses is conceived."[62] The workings of the soul are felt throughout the human body, which reveals how it is connected to the body.[63]

Since a person is regarded as body and soul, as a psychosomatic unity and existence, the noetic nature of the mind renders possible its presence throughout the entire body;[64] the soul is present in the entire human body, as the mind is united and pervades the whole body.[65]

The union of the soul and mind with the body is revealed and discerned through the sensory organs.[66] Although humans' sensory organs are diverse, the mind acts through all the organs, maintaining its unified power.[67]

Gregory holds that souls and bodies are created simultaneously: "In each act of creation, no preference is given to one over the other; neither to the soul before the body nor to the body before the soul, as if humanity were not in unity with itself, apportioning each differently across time,"[68] because the soul and the body are built by God at the same time and coexist. "The creation of both the soul and the body, which together constitute the basis of existence, implies that a person, an entity composed of soul and body, possesses a unified and common nature from the outset. It is as though he does not precede or follow himself; neither his physical aspect is prior within him, nor is the spiritual subsequent."[69]

The mystical union of the soul with the body takes place at the time of the conception of the embryo. The soul is innate with the body and is not installed in it.[70] From the very moment of conception, the soul connects the body,[71] allowing the soul to perceive the world through the body's senses and sensory organs.[72]

A human's identity is defined by the organic unity of soul and body, while his existence, right from the first moment of his creation, and the perpetuation of his life are determined by and dependent on his relationship with God as a relationship of communion between Creator and creation. This relationship of communion between God and human, when lasting and solid, gives humans the prospect of a continuous journey towards salvation and immortality.[73]

According to the anthropology of Gregory of Nyssa, the union of the tangible world with the body and that of the mental world with the soul is accomplished within humans. When humans reflect on their inner world, they becomes aware of the soul.[74] The rational human depicts God within whom the words of beings exist. Their rationality also derives from the fact that humans summarize the words of beings in the image of God's Word.[75]

The rational part of human nature relates to the emotional, the desiring, and the sensory. Thus, the mental logical power is connected with the senses, which are inherent and interact with the body. When the logical part prevails over the emotional and the desiring part, through the proper management of the senses, the rational prevails over the irrational and is then eagerly led through virtue to perfection.[76]

Since the mind and the soul are considered to be of an unseen and noetic nature, the mind permeates the whole human existence.[77] Thus, the mind is the distinctive characteristic of humans and the major part of their soul, and the rational soul is grasped by the manifestation of its actions, which is the unique and special characteristic that distinguishes humans from the rest of creation.[78]

Regarding the soul, according to Symeon the New Theologian, the soul of humans is a "logical image of the Word;" the soul of a person as "in the image" of God's Word refers to the original, with the unified existence of the soul, the mind, and the reason with the "three in one unity" of the Father, the Son and the Word and the Holy Spirit.[79]

The rational, mental, and immortal tripartite soul, as an image, bears the characteristics of the divine idioms and a sequence of relations as in the triune God. The mind, soul, and immanent reason are equal, consubstantial, of one essence and nature. Soul, mind, and reason, as "intrinsic" distinct manifestations of the one common essence and nature, share the common nature and essence according to the prototype of the Triune God.[80]

Gregory of Nyssa, approaching the "in the image" of the soul, parallels it with the three hypostases of the Holy Trinity as "in the image" of the Triune God.[81] He divides the "in the image" of the soul into three parts, following the model of the three hypostases of the Holy Trinity, namely the unbegotten of the soul, the genesis of the reason, and the origin of the spirit,[82] so that the distinction in the tripartite of the soul indicates the mystery of the Holy Trinity.[83]

Upon the union of the three hypostases of the soul, the unified energy emerges, as an image of the common and indivisible divine energy.[84] God, on the creation of humans and their soul, proclaims to the world the mystery of the Holy Trinity.[85] Therefore, humans, being created "in the image" and "in the likeness" of God, become a potential partaker of the divine actions.

According to patristic anthropology, humanity is considered a "medium" and "frontier" between the spiritual and the physical creation, which harmonizes and unites objects.[86] Maximus the Confessor holds that humans were created last of all creation and placed in the middle of the world as a natural link to intercede and bring together the extremes, leading to unity and starting from himself. Consequently, man's destiny in the creation and the world is the role of the mediator so that unity can be achieved.[87]

The role of the mediator involves two levels and directions. Initially, once the unity of human existence is achieved, the ascending spiritual journey can be carried out, aiming at a union with God. At the next level, as a mediator, they are called upon to play a mediating role in bridging any divide or disruption that prevails in the world. Humans as a medium contribute to the unity of the mental with the tangible through the mutual complementation and interpenetration of the parts which takes place by means of their personal integration and unity with God.[88] As for the physical part, the intent is the proper use of the body and its passions.[89]

Humans as communicants of the Divine Economy unify their existence, which has been disrupted by sin, so that through the exercise of virtue, in freedom, along with the proper use of their self-authority, they are healed and regenerated by the restorative grace in order to become children of God by grace and the Gospel exhortation "be holy, for I am holy"[90] is fulfilled.

## THE "DEIFIED" MAN

Before the creation of humans by God, the cosmic, insensate, and physical elements of the world preceded, and then sentient life was created. Humans are formed only after the vegetation and the irrational creatures to include, upon their creation, the entirety of the world's life force. Thus, they encompass the traits of both the material-inanimate and animate animal world, yet they are endowed with the intellectual and rational as their unique characteristic and as the medium of creation's communion with God. This distinct characteristic is reinforced by the breath which God instilled in their face, so that they possess the characteristics of the microcosm and so that the visible relates to the invisible, the mind with the senses, the earthly with the heavenly, the material with the immaterial, the mundane with the eternal, which as an image and likeness of God can be deified.[91]

Humans were called upon to gather within themselves all of creation, so that through the union with God they can impart this deified status to the creation.[92] Their purpose was the permanent communion with God, so they could transform the entire earth into a paradise, eliminating the distances of the physical world and uniting heaven with earth. By transcending the boundaries of the tangible, they can then emerge into the mental world, to include the mental and the tangible world within themselves. Humans' destiny was to completely devote themselves to God, with love, displaying the whole universe that is gathered within the human being. Then, as a result, God bestows Sonship on humans, who will be by Grace what God holds by nature.[93] However, this union takes place within the premises of the Church, which is sanctified by the sacraments that serve as conduits of Grace.

Through the cleansing of a person from its passions, upon the enlightenment, humans become aware of the reason of beings. During theosis-deification, human's grace-filled likeness and union to God is realized.[94] The entire human is sanctified in soul and body and becomes God by grace.

A human is formed as "a mortal being created from the earth" with reference to the visible creation and honored "by grace immortal" with reference to its invisible creation so that it can be observed and praise God.[95] The sentient and mental synthesis of humans as a psychosomatic unity enables them to behold, through the mental soul, the uncreated light, the glory of God and to participate in the mental world.[96]

The created human, consequently, possesses the possibility to cultivate a personal relationship and communion with the uncreated Trinitarian God in terms of the mental soul, "the human within," to behold the divine light, to speak with God in a person-to-person basis thanks to the fact that he has a mental soul, to behold the mental sun-God, who is Christ Himself.[97] The breath of life that humans received from God creates a kinship, a closeness between humans and God. Humans, as living creatures, are destined to become "deified beings," who have the vocation to become gods.[98]

Only humans represent the Triune God as having a mind, reason, and spirit. For the sentient and irrational animals possess only a vital spirit, "having neither mind nor reason," whereas the super-sentient, such as the mental and rational angels and archangels, have a mind and reason, but not a life-giving spirit, "neither do they have a living body." Humans are uniquely created "in the image" of the triune nature and possess a mind, a reason, and a spirit that gives life to the body, "having a living body,"[99] that is receptive to divine enlightenment.

Apart from the "in the image," according to Symeon the New Theologian, "in the likeness" identifies the light, the glory of God, and the divine attributes.[100] Humans in Paradise donned "repression" as a vesture, the clothing of divine glory, in both soul and body and wearing it, stood in the glory of God arising from communion with God prior to the Fall.[101] This garment of glory signifies the possibility of being made "in the likeness" through the glorious light of the virtues and the active union with God by grace.

The uncreated grace of God grants humans a divine form and unites them with God as a single psychosomatic existence, and the created humans attain a state beyond nature and time, and the deified becomes the world's light.[102]

Despite the ontological gap between the created human and the uncreated God, the possibility of the theosis-deification of the created human involves his entire psychosomatic nature through the immersion of divine light as in the Transfiguration of the Lord, maintaining the personal identity characteristics of each one as a unique event of theosis-deification.

According to Dionysius the Areopagite, the purpose of humans' creation is "the assimilation and union with God, as much as possible."[103] While Maximus the Confessor states that God created humans to partake of his own immortality and become a communicant of the divine nature and be like Him "in theosis-deification by Grace."[104] The requisites of theosis-deification are found in the "in the image and likeness" of humans to God, which is offered as a gift of God to humans.[105]

During the sonship and union, by grace, humans become a triune "God by position, εἰς θέσει Θεός," of body and soul "and from whom he received the

Holy Spirit" as a divine begotten of the Holy Spirit "in the image and likeness of God."[106]

The indwelling of God in the heart of the human, as the psychosomatic center, and the union by grace, through the enlightenment of the Holy Spirit, applies to the entire humanity, body, and soul, and from being dual becomes threefold through the indwelling of God.[107] During the incarnate enlightenment, the human mind transcends the bodily senses and, in light, considers God through his spiritual senses.[108]

Upon union with God, the totality of humans becomes God "by grace," and the regenerate human is united with God by grace, through the gift of God, as brothers of Christ and sons of God in the same way that the persons of the Holy Trinity are united by nature.[109]

The union of humans with God according to the theory of the divine light and through the participation of the sacred Mysteries applies to the whole human as a psychosomatic unity, so that the union with God by grace may take place in the twofold nature of human, soul, and body.

The created human maintains his referentiality to the uncreated archetype which enables him, being in His image, to resemble God, Jesus Christ, and by His uncreated grace to become God by Grace, to be deified.

Gregory the Theologian emphasizes the uniqueness of humans, describing them as immortal and mortal, modest and elevated, humble and great, worldly and heavenly, having earthly and carnal but also divine and spiritual characteristics, as "Christ must be buried with us, Christ must rise together, inherit together so that he becomes sons of God."[110]

## THE "NEW HUMANITY" IN JESUS CHRIST

Following the Fall, humans were not capable of overcoming "the limitations of their nature" and the consequences of the Fall on their own. The recovery and the overcoming of the fallen nature are brought about by the incarnation of the Son and Word of God and the synergy of the Triune God within the body of the Church.[111]

Yet the Economy of God provided for the regeneration of humans, through the incarnation of the Son and Word of God in the person of Jesus Christ from the fall that preceded it. According to Gregory of Nyssa, in the first creation humans were formed but through disobedience they were corrupted; in the second creation they were restored into a new creation through Christ. "We have known the twofold nature of our creation, the first in which we were created and the second in which we were restored, but we wouldn't need the second creation, unless we had lost the first through disobedience. Therefore, having been made old and destroyed, a new creation was born again in Christ."[112] The entire human race communes with the divinity and, consequently, is healed and regenerated through Jesus Christ.[113]

Under the Divine Economy, with the incarnation of God and His Word, the "new creation"[114] is established, the course of a person from life to death is reversed, since in Christ he "passed from death to life."[115] And within the worldly space and time his life is "encapsulated with Christ in God."[116]

Moreover, according to Gregory the Theologian, the eternal Word of God takes on human nature, and from this communion and union of God with the whole of humanity, humans attain a second communion with God, greater than the first, which results in the theosis-deification of the assumed human nature.[117]

Through Christ's Economy, human nature is renewed, and every human being can be perfected and deified. He casts off the old, fallen human and puts on "the new one, who is being renewed in the image of the Creator,"[118] to "renew the spirit of our minds."[119] According to Basil the Great, "the economy of God and our Savior, as far as a person is concerned, is the revocation of the fall and the return in the presence of God after the separation caused by disobedience."[120] In the words of Symeon the New Theologian, "in soul and body, the human built in the image of God and is God worthy" by grace.[121]

Athanasius the Great considers that God's ultimate benevolence towards humans is the incarnation of the Son and Word of God,[122] who as "mediator of God and men serves as the only philanthropist."[123] Christ grants eternal life because He is the life and resurrection of the world, and humans are freed from death by the power of Christ's Resurrection.[124]

In the face of Christ, the new Adam, the perfect human, who is united with God and recapitulates creation, is revealed. As a result, Christians as members of the body of Christ of the Church become a universal human, like Christ, and a Christ by grace.[125] The body of Christ like a new root offers its fruits and shoots, the incorruption and eternal life.[126]

The return and recovery from disobedience, which caused the fall of humans, came about through the plan of the Divine Economy by the incarnation of the Son and Word of God, so that humans could be saved and receive the sonship through the imitation of Christ.[127]

The purpose of the incarnation of God's Word is the salvation, the regeneration of the human race, the liberation of humans, and the restoration of their self-authority by being liberated from sin through His passion. Furthermore, the purpose of the incarnation is the theory of divine glory and human's theosis-deification by grace "in order for God to create the man."[128]

The renewal of a human as someone created in the image of God is realized by Christ and is transformed into the believer by the Holy Spirit.[129] The new human in Christ is spiritual because he partakes of the Holy Spirit's grace.[130] In the words of Palamas, "The spiritual person is born of three things: the grace of the heavenly Spirit, the rational soul, and the earthly body."[131] The perfect human is composed of the soul, which the Spirit of God has received and connected to the flesh created in His image.[132]

Through the incarnation of the Son and Word of God, the theosis-deification of humans is accomplished, as He assumed and deified human nature.[133] Christ opens the way of salvation and theosis-deification.

Just as Christ remained God, although he was born as a human, so the soul of the spiritual person is reborn in his union with the God-Man Jesus and bears his image.[134] God's Word, by His incarnation, "has mysteriously renewed our whole nature."[135]

The renewal of humanity in Christ frees us from the Fall. It restores the unnatural back to the natural and offers the possibility of theosis-deification and union with God to achieve the supernatural.[136] A human enters a "lightning-like apathy" upon his union with Christ, and the fire of divinity shines within his mind.[137]

Through the Ascension of Christ, human nature is led upward, beyond the pre-Fall state, into the blessedness of the kingdom of God, and Christ becomes "Father of the age to come."[138] Thus, the God-Human as the "new human" abolishes the old, unifies the separated, and unites humanity in His substance, "creating the two in himself into one new human."[139]

Upon the renewal in Christ, God's Grace sanctifies people "wholly"[140] as a single psychosomatic unity. The energy of the renewing grace sanctifies the soul as well as the body, to become a temple of the Holy Spirit.[141] The new human is directly related to the moral and spiritual life of the believer, since he is asked to abandon "the previous conduct, the former human,"[142] to cast out the former human of passions.[143]

And to put on, with awareness, the new human of the renewing virtue.[144] Humans receive a newly acquired value in the Church, without discrimination, as a link of unity, equality, and love. The new human is built within the Church,[145] who can be perfected "into a perfect human, to the measure of the age of the coming of Christ."[146]

Christian anthropology, as reflected in Pauline theology, is the theology of the "new human" due to the Divine Economy as opposed to the "former human." The differentiation of the former and the new human is initially sacramental, beginning with the Sacrament of Baptism and concluding with the other sacraments. Baptism acts as a "bath of regeneration and renewal of the Holy Spirit."[147] The entirety of humanity puts on Christ, gets renewed, and participates sacramentally in His death and Resurrection.[148]

The incarnate Word of God became the Savior, the path, and the truth of the Church's body, carrying out the duties of the royal, prophetic, and high priestly office, and every member of the Church became a communicant of the divine nature of the mystical body of Christ.[149]

According to Professor Skouteris, "the threefold of Paul's theology on the new man, the sacramental sphere, the personal spiritual life and the ecclesiastical society, constitutes a solid basis for the documentation of a comprehensive Christian anthropology."[150]

The "New Human" is reconstructed, renewed, and deified through Jesus Christ in the Church, participating in His incarnation, sufferings, crucifixion, Resurrection and Ascension. The renewed human becomes a partaker of the divine nature of the mystical body of Christ, through the Sacraments of the Church.

## NOTES

1. John Zizioulas, Metropolitan of Pergamon, "Man and the Environment: An Orthodox Theological Perspective." In *Sciences, Cutting-Edge Technologies and Orthodoxy*, Holy Synod of the Church of Greece, pp. 275–284 (Athens: 2002). Chrysostomos Savvatos, metropolitan bishop of Messenia, "Man and the Evolution of Modern Technology. A Challenge for Orthodox Theology," Holy Metropolis of Messinia (Kalamata: 2019), p. 26.

2. Chrysostomos Savvatos, metropolitan bishop of Messenia, "Man and the Evolution of Modern Technology. A Challenge for Orthodox Theology: The Technology Phenomenon," I, *Theology* 90, no. 3 (2019): 34, 35.

3. Chrysostomos Savvatos, metropolitan bishop of Messenia, "Man and the Evolution of Modern Technology. A Challenge for Orthodox Theology," p. 27.

4. Anastasios Yannoulatos, Archbishop of Tirana, Durrës and All Albania, "Orthodox Views About the Development of Science and Technology," *Welcome Address*, November 15, 2020 [https://orthodoxalbania.org/2020/el/2020/11/15/%ce%bf%cf%81%ce%b8%cf%8c%ce%b4%ce%bf%ce%be%ce%b5%cf%82-%ce%b1%cf%80%cf%8c%cf%88%ce%b5%ce%b9%cf%82-%cf%83%cf%87%ce%b5%cf%84%-ce%b9%ce%ba%ce%ac—%ce%bc%ce%b5—%cf%84%ce%b7%ce%bd-%ce%b1%ce%bd%ce%ac%cf%80/ (17/01/2021)].

5. Georgios Lekkas, Archpriest, Counsellor at the Representation Office of the Church of Greece to the E.U., "Artificial Intelligence and an Applied Ethics. But What Kind of Ethics?" Brussels, January 21, 2019 [https://ec.europa.eu/futurium/en/european-ai-alliance/artificial-intelligence-and-applied-ethics-what-kind-ethics (19/12/2020)].

6. Georgios Lekkas, Archpriest, Counsellor at the Representation Office of the Church of Greece to the E.U. "Artificial Intelligence and an Applied Ethics. But What Kind of Ethics?"

7. Official Documents of the Holy and Great Council of the Orthodox Church, *Encyclical V 11. The Church in the Face of Globalization, Extreme Violence and Migration § 16*, Crete, June 2016 [https://www.holycouncil.org/-/encyclical-holy-council (19-11-2020)]. Official Documents of the Holy and Great Council of the Orthodox Church, *Message, § 7*, Crete, June 2016 [https://www.holycouncil.org/-/message

(11-11-2020)]. The term "Church" according to the ancient Greek classical usage and etymology comes from the verb *ek-kaleo-o,* which means "to gather, welcome, convene, invite," Plato originally mentions it, as well as other philosophers, such as Aristotle, Polybius, and Thucydides. Aristotle, *Politics* 3,9 (14), 4. 7,2. Plato, *Gorgias* 456B. *Politeia*, VI,492B. Thucydides, *History of the Peloponnesian War*, 1,31,87. Polybius, *Histories*, 5,74. Ioannis Karmiris, *Orthodox Ecclesiology, Dogmatics Part E'*, Athens 1973, p. 11.

8. Symeon the New Theologian, *Discourses vol. 1*, 163–168, ed. J. Darrouzes, *Sources Chrétiennes (SC)* 122 Paris 1980, p. 108.

9. Athanasius the Great, *To the Antiochian Lord*, *PG* 28, 608A. John Damascene, *An Exact Exposition of the Orthodox Faith*, *PG* 94, 921A. Athenagoras, *On the Resurrection of the Dead*, *PG* 6, 1000B. Gregory of Nyssa, *Against Apollinarius*, 54, *PG* 45, 1253C–1256A. Symeon the New Theologian, *The Ethical Discourses vol. 1*, a', 19–25, *SC* 122, p. 174. Gregory of Nyssa, *On the Making of Man C'*, *PG* 44, 133CD. Gregory of Nyssa, *On the Making of Man*, *GCF* 5, Patristic Publications, "Gregory Palamas" (Thessaloniki: 1987), p. 56. Genesis 1,26–28.

10. George Mantzaridis, *Christian Ethics II, Man and God, Man and Fellow Man, Existential and Bioethic Propositions and Perspectives* (Pournara, Thessaloniki: 2004), p. 495.

11. Symeon the New Theologian, *The Ethical Discourses vol. 1*, g', 4–5, *SC* 122, p. 236. Gregory of Nyssa, *On the Making of Man*, *GCF* 5 (Thessaloniki: 1987), p. 122.

12. John Chrysostom, *Psalm 144* 1, *PG* 55, 465. Symeon the New Theologian, *Hymn 53*, 102–104, *SC* 196, p. 218. *The Ethical Discourses vol. 1*, j', 69–70, *SC* 122, p. 256.

13. John Damascene, *Dialectic*, *PG* 94, 609A. Symeon the New Theologian, *Hymn 25*, 63, *SC* 174, p. 260.

14. Symeon the New Theologian, *Hymn 23*, 448–455, *SC* 196, p. 218. *Hymn 53*, 112–131, *SC* 196, p. 220.

15. Basil the Great, *Isaiah 1*, 13, *PG* 30, 140, Ps. 32, 6, *PG* 29, 337.

16. John Chrysostom, *On Genesis, Homily*, 14, 5, *PG* 53, 117.

17. Gregory of Nazianzus, *Discourse*, 38, 11, *PG* 36, 317. Ioannis Karmiris, *Synopsis of the Doctrine of the Orthodox Catholic Church* (Athens: 1960), p. 27.

18. John Chrysostom, *On the Ambiguities of the Old Testament*, 2, 5, *PG* 56, 182. John Damascene, *On Christ's Two Wills*, 30, *PG* 95, 168.

19. Gregory of Nyssa, *On the Making of Man C'*, *PG* 44, 133CD. Nikolaos Xionis, *Man and Society, The Creation of the World and Human Nature* (Athens: Ennoia, 2020), p. 192.

20. Gregory of Nyssa, *On the Making of Man D'*, *PG* 44, 136B.

21. Gregory of Nyssa, *Catechetical Discourse*, 5, J. Srawley, pp. 26, 30, *On the Soul and Resurrection*, *PG* 46, 101B, *On the Making of Man D'*, *PG* 44, 136B, N. Xionis, *Man and Society*, p. 196.

22. John Chrysostom, *On Genesis, Homily*, 8,2, *PG* 53,71.

23. John Chrysostom, *On "My Grace Is Sufficient for You,"* 2, *PG* 59, 509.

24. Gregory of Nyssa, *On the Making of Man H'*, *PG* 44, 144B. N. Xionis, *Man and Society*, p. 225.

25. John Chrysostom, *Goths Reading, and a Gothic Elder*, spoken in the Church of Paul, 8, *PG* 63, 508–510.

26. John Chrysostom, *Homily on Romans* 8, 2, *PG* 60, 510. John Damascene, *An Exact Exposition of the Orthodox Faith*, 2, 12, *PG* 94, 320. Nicholas Kabasilas, *On the Life in Christ, Discourse 5*, *PG* 150, 638. Dimitrios Tselengidis, *Grace and Freedom in the Patristic Tradition of the Eleventh Century*, (Pournara, Thessaloniki: 1987), pp. 40, 41.

27. N. Xionis, *Man and Society*, p. 187. John Damascene, *Against Those Who Decry the Holy Images C'*, *PG* 94, 1340D–1341. *Genesis* 1,26–28. Alberca I. Escribano, "Die Entdeckung des Inneren Menschen," to *Gregor von Nyssa und die Philosophie, II Internationales Kolloquium über Gregor von Nyssa*, Freckenhost bei Münster 1972 (Leiden: E. J. Brill, 1976), pp. 43–58. Gregory of Nazianzus, *Theological Oration V*, Discourse LA, *PG* 36, 169A–172A. Nikolaos Braciotos, *Anthropology of the Old Testament. The Man as Divine Creation* (Athens: 1989), pp. 262–267.

28. Theodoros Zisis, *Man and the World Under the Economy of God According to the Holy Chrysostom*, Patriarchal Foundation for Patristic Studies (Thessaloniki: 1971), pp. 77,78. John Chrysostom, *Psalm 48*, 7, *PG* 55, 233, *Homily on Ephesians* 21, 4, *PG* 62, 154, *Homily on First Thessalonians* 10, 3, *PG* 62, 458.

29. Chrysostomos Savvatos, metropolitan bishop of Messenia, "Man and the Evolution of Modern Technology. A Challenge for Orthodox Theology," 24. Irenaeus of Lyons, *Against Heresies*, 5, 6, 1, *PG* 7, 1158AC.

30. Ioannis Karmiris, *Christian Ethics*, vol. A' (Athens 1967), p. 41.

31. John Damascene, *An Exact Exposition of the Orthodox Faith II*, 12 *PG* 94, 920.

32. Gregory of Nyssa, *Let Us Make Man in Our Image and Likeness*, *PG* 44, 273B. John Chrysostom, *On Genesis Discourse* 2, 3, *PG* 54, 590, *On Genesis Homily* 9, 3, *PG* 53, 78. Panayiotis Nellas, *Devinely Inspired Animal, Perspectives for an Orthodox Understanding of Man*, (Athens: Epopeia, 1979), p. 19. Ioannis Karavidopoulos, *God's Icon and "in the Image" of God by the Apostle Paul. The Christological Foundations of Pauline Anthropology* (Thessaloniki: 1964), p. 21. Louis Meyer, *St. J. Chrysostome, Maître de Perfection Chrétienne* (Paris: Gabriel Beauchesne et Ses Fils, 1933), pp. 207.

33. Gregory of Nyssa, *Let Us Make Man in Our Image and Likeness*, *PG* 44, 273A.

34. Jean Daniélou, *Platonisme et Théologie Mystique. Essai sur la Doctrine Spirituelle de Saint Grégoire de Nysse* (Paris: 1944), p. 52. Elias Moutsoulas, *The Incarnation of the Word and the Theosis-Deification of Man according to the Teaching of Gregory of Nyssa* (Athens: 1965), pp. 68, 69.

35. Gregory of Nyssa, *Let Us Make Man in Our Image and Likeness*, *PG* 44, 273A. Nikolaos Matsoukas, *Dogmatic and Symbolic Theology B'* (Pournara, Thessaloniki: 2001), pp. 195, 196. Archimandrite Aristarchos Gkrekas, *Patristic Literature and Theology I* (Thessaloniki: Ostracon Publishing p.c., 2020), pp. 98–100.

36. Gregory of Nyssa, *On the Making of Man*, *PG* 44, 184B. Gregoire De Nysse, *La Création de l'Homme*, *SC* 6, J. Laplace (Paris-Lyon: 1943), pp. 157, 184b.

37. Gregory of Nyssa, *Catechetical Discourse*, *PG* 45, 24C.

38. Gregory of Nyssa, *Catechetical Discourse*, *PG* 45, 77A.

39. Gregory of Nyssa, *Catechetical Discourse*, *PG* 45, 77CD.

40. Pachomios Rusanos, "To Dionysius." Ioannis Karmiris, *Pachomios Rusanos and the Unpublished Dogmatic and Other Works of Him*, Texte und Forschungen zur byzantinischneugriechischen Philologie, 14+ (Athens: 1935), pp. 174, 176. Athanasius the Great, *Discourse on the Incarnation of the Word and Through the Body to Ourselves Epiphany*, *PG* 25, 95–198. Vlasios Feidas, *Ecclesiastical History*, vol. A' (Athens: 1997), pp. 556–560. Georgios Rallis, *Constitution of the Divine and Sacred Canons of the Holy and Praised Apostles and of the Sacred Ecumenical and Local Councils and of the Local Holy Fathers*, from the printing press of G. Hartophylakos, ed. Potlis Michail, Athens 1852–1859 C′, 561 [https://anemi.lib.uoc. gr/metadata/9/3/c/metadata-01–0000374.tkl (10/02/2021). John Chrysostom, *Homily on First Corinthians*, 17, 3, *PG* 61, 143. *To Stageirion*, 1, 2, *PG* 47, 428. *Homily on Romans*, 11, 2, *PG* 60, 486. Panagiotis Christou, *The Teachings on Man of Theophilus of Antioch* (Thessaloniki: 1957), p. 15.

41. Gregory Palamas, *Homily*, 16, *PG* 151, 196 A.

42. George Mantzaridis, *Christian Ethics II, Man and God, Man and Fellow Man, Existential and Bioethic Propositions and Perspectives*, p. 636.

43. Pachomios Rusanos, "To Dionysius." Ioannis Karmiris, *Pachomius Rusanos and the Unpublished Dogmatic and Other Works of Him*, p. 186. John Chrysostom, *On Genesis, Homily* 16, 5, *PG* 53, 132. *Homily on Romans* 9, 16. Legrand Philippe-Ernest, *St. J. Chrysostome, Les Moralistes Chrétiens*, J. Gabalda (Paris: 1924), p. 40, 41.

44. Pachomios Rusanos, "To Dionysius." Ioannis Karmiris, *Pachomius Rusanos and the Unpublished Dogmatic and Other Works of Him*, pp. 187–195.

45. Gregory Palamas, *On the Divine Methexis*, 19, ed. P. Christou. Gregory Palamas, *Writings*, vol. 2 (Thessaloniki: 1970), p.153. A' Timothy, 6, 16.

46. Gregory of Nyssa, *On the Making of Man*, *GCF* 5, p. 32.

47. Genesis 2,7. Olympia Papadopoulou-Tsanana, *The Anthropology of Basil the Great*, ed. Patriarchal Foundation for Patristic Studies (Thessaloniki: 1970), p. 35.

48. Genesis 1,26.

49. Genesis 2,7.

50. Symeon the New Theologian, *The Ethical Discourses vol. 13*, 131–133, *SC* 129, p. 410. *The Discourses vol. 1*, 184–185, *SC* 122, p. 110.

51. John Damascene, *An Exact Exposition of the Orthodox Faith*, *PG* 94, 853A and *PG* 94, 1049. Clement of Alexandria, *Stromata*, 6, 15, *PG* 9, 516B, 13. Basil the Great, *Homily C', On the Care of Thee*, *PG* 31, 215.

52. Basil the Great, *Isaiah*, *PG* 30, 140.

53. Basil the Great, *On the Care of Thee*, *PG* 31, 212B.

54. Alexandros Marsselos, "Basil the Great as a Physician," Vassilias, festive volume on the completion of 1,600 years since the death of Basil the Great, Holy Metropolis of Thessaloniki (Thessaloniki: 1979), p. 193.

55. Gregory of Nyssa, *Let Us Make Man in Our Image and Likeness*, *PG* 44, 280D.

56. Gregory of Nyssa, *Catechetical Discourse*, *PG* 45, 25D.

57. Gregory of Nyssa, *Catechetical Discourse*, *PG* 45, 25D.

58. Ioannis Plexidas, *The Anthropology of Evil, The Problem of Theodicy and Its Anthropological Presuppositions in the Thought of Gregory of Nyssa*

(Thessaloniki 2010), p. 195. Gregory of Nyssa, *On the Soul and the Resurrection*, *PG* 46, 12–166. Gregory of Nyssa, *Catechetical Discourse*, *PG* 45, 93A.

59. Gregory of Nyssa, *On the Soul and the Resurrection*, *PG* 44, 232A.

60. Gregory Palamas, *Chapters*, 61, *PG* 150, 1165A.

61. Panagiotis Christou, *Gregory Palamas' Writings*, vol. A' (Thessaloniki: 1962), p. 85.

62. Gregory of Nyssa, *The Complete Works I. Dogmatic Treatises A',* Series Greek Church Fathers, trans. Athenagoros Kokkinakis, Archbishop of Thyateira and Great Britain, *GCF*, Patristic Publications "Gregory Palamas" (Thessaloniki: 1979), p. 89. Gregory of Nyssa, *On the Soul and the Resurrection*, *PG* 46, 29B.

63. Gregory of Nyssa, *On the Soul and the Resurrection*, *PG* 46, 44BC.

64. Gregoire De Nysse, *La Création de l'Homme,* pp. 131, 161c. Gregory of Nyssa, *On the Making of Man*, *PG* 44, 161C. Georgios Skaltsas, "Man as a Mirror of the Eschaton according to Saint Gregory of Nyssa," *Synaxis* 59 (1996): 45–49.

65. Nikolaos Matsoukas, *The Problem of Evil* (Kyriakidis, Thessaloniki: 2009), pp. 84, 86.

66. Gregoire De Nysse, *La Création de l'Homme*, pp. 99, 137d–140a. Gregory of Nyssa, *On the Making of Man*, *PG* 44, 137D, 140A.

67. Gregoire De Nysse, *La Création de l'Homme*, p. 99, 140a. Gregory of Nyssa, *On the Making of Man*, *PG* 44, 140A.

68. Gregory of Nyssa, *On the Making of Man*, Writings, *GCF* 5 (Thessaloniki: 1987), p. 208.

69. Gregory of Nyssa, *On the Making of Man*, *PG* 44, 233–240. Georgios Skaltsas, "Man as a Mirror of the Eschaton according to Saint Gregory of Nyssa," p. 123. Ioannis Plexidas, *The Anthropology of Evil, the Problem of Theodicy and Its Anthropological Presuppositions in the Thought of Gregory of Nyssa*, p. 129.

70. John Damascene, *An Exact Exposition of the Orthodox Faith*, 13, *PG* 94, 853A.

71. Skaltsas, "Man as a Mirror of the Eschaton according to Saint Gregory of Nyssa," p. 123.

72. Skaltsas, "Man as a Mirror of the Eschaton according to Saint Gregory of Nyssa," p. 99.

73. Philipisious, 3:13. Gregory of Nyssa, *On the Song of Songs*, *PG* 44, 772A. Chrysostomos Savvatos, metropolitan bishop of Messenia, "Man and the Evolution of Modern Technology. A Challenge for Orthodox Theology," pp. 16, 17.

74. Gregory of Nyssa, *On the Soul and the Resurrection*, *PG* 46, 28C. Gkrekas Aristarchos Archim., *Patristic Literature and Theology I*, p. 102.

75. George Mantzaridis, *Christian Ethics II*, p. 41. Athanasius the Great, *Discourse on the Incarnation of the Word* 5, *PG* 20, 101B. Symeon the New Theologian, *Hymn* 44, 30–37, *SC* 196, p. 72. *Ethical vol.* 3, 358–362, *SC* 122, p. 416.

76. Gregory of Nyssa, *On the Making of Man*, 315 IH', *PG* 44, 192BC. Gregory of Nyssa, *On the Soul and the Resurrection*, *PG* 46, 60D–65C. John Damascene, *On Two Wills*, *PG* 95, 148.

77. Nikolaos Matsoukas, *The Problem of Evil. Essay on Patristic Theology*, ( Pournara, Thessaloniki: 1986), pp. 84, 86.

78. Olympia Papadopoulou-Tsanana, *The Anthropology of Basil the Great*, p. 35. Gregory of Nyssa, *The Complete Works I.* Dogmatic Treatises A', Series Greek Church Fathers, trans. Athenagoros Kokkinakis, Archbishop of Thyateira and Great Britain, ed. GCF Patristic Publications "Gregory Palamas" (Thessaloniki: 1979), p. 80.

79. Symeon the New Theologian, *Hymn* 44, 30–54, *SC* 196, p. 2. *The Ethical Discourses* vol. 3, 358–362, *SC* 122, pp. 72–74, 416.

80. Maximus the Confessor, *Letters XVI, To Him on Slander and Redemption by Divine Assistance*, *PG* 91, 557D–560A. John Damascene, *Dialectic*, *PG* 94, 616AB. Symeon the New Theologian, *The Discourses* vol. 2, 89, *SC* 118–122, 138.

81. Gregory of Nyssa, *On What It Is "in the Image" and "in the Likeness" of God*, *PG* 44, 1336AB.

82. Nikolaos Matsoukas, *Dogmatic and Symbolic Theology, C', Recapitulation and Agathotopia.* Exposition of the ecumenical character of Christian teaching, ed. Pournara, Thessaloniki 1997, p. 60. Gregory of Nyssa, *On What It Is "in the Image" and "in the Likeness" of God*, *PG* 44, 1336A.

83. Gregory of Nyssa, *On What It Is "in the Image" and "in the Likeness" of God*, *PG* 44, 1344C.

84. Gregory of Nyssa, *On What It Is "in the Image" and "in the Likeness" of God*, *PG* 44, 1344AB.

85. Gregory of Nyssa, *On What It Is "in the Image" and "in the Likeness" of God*, *PG* 44, 1340CD.

86. Chrysostomos Savvatos, metropolitan bishop of Messenia, "Man and the Evolution of Modern Technology. A Challenge for Orthodox Theology," p. 19.

87. Maximus the Confessor, *On Various Questions*, *PG* 91, 1305B. Aristarchos Gkrekas Archim, *The Church as a Place of Communion between God, Man and the World*, (Thessaloniki: Ostracon Publishing p.c., 2014), pp. 29, 30. John Chrysostom, *On First Timothy, Homily* 7, 2, *PG* 62, 536.

88. Maximus the Confessor, *Mystagogy*, ed. Apostoliki Diakonia, Athens 1973, p. 116. Aristarchos Gkrekas Archim., *The Church as a Place of Communion between God, Man and the World*, p. 30. Symeon the New Theologian, *Hymn 53*, 116–119, *SC* 196, p. 220. Anthimos Gazis, *Interpretation of Genesis*, *PG* 87, 128C.

89. Xionis, *Man and Society*, p. 225. John Damascene, *Against Manichaeus* 14, *PG* 94, 1517C 1520A. *On Christ's Two Wills*, *PG* 95, 149A. *Bible of Questions*, *PG* 90, 1109D–112A.

90. Peter A' 1,16.

91. Xionis, *Man and Society*, pp. 192, 193. Gregory of Nazianzus, *On the Epiphany*, Discourse LH, 11, *PG* 36, 321C–324B. Andreas Theodorou, *The Perceptions of Saint Gregory of Nazianzus Regarding the Human Body*, SYSTA KA' (Athens: 1975), p. 9. Nikolaos Xexakis, *Orthodox Dogmatics C'. The Doctrine of Creation* (Athens: Ennoia, 2009), p. 125. Chrysostomos Savvatos, metropolitan bishop of Messenia, "Man and the Evolution of Modern Technology. A Challenge for Orthodox Theology," p. 19. Nikolaos Matsoukas, *Dogmatic and Symbolic Theology B', Exposition of the Orthodox Faith* ( Kyriakidi, Thessaloniki: 2016), p. 182. Kallistos Ware, Bishop of Diokleia, "The Unity of the Humanperson: The Body-Soul Relationship

in Orthodox Theology," Sciences, Cutting Edge Technologies and Orthodoxy, International Scientific Conference, October 4–8, 2000, Holy Synod of the Church of Greece (Athens: 2002), pp. 338–342. Wolfhart Pannenberg, *Was ist der Mensch? Die Anthropologie der Gegenwart im Lichte der Theologie*, Vandenhoeck & Ruprecht (Göttingen: 1976), p. 5.

92. Vladimir Lossky, *The Mystical Theology of the Eastern Church* (Thessaloniki: 1986), p. 123.

93. Maximus the Confessor, *On Various Questions*, *PG* 91, 1308. Aristarchus Grekas Archim., *The Church as a Place of Communion between God, Man and the World*, pp. 30, 31.

94. Nicholas Kabasilas, *On the Life in Christ*, 6, *PG* 150, 680AB.

95. Gregory Palamas, *Homily XI, On the Precious and Life-Giving Cross*, 18, *PG* 151, 137B. Symeon the New Theologian, *Hymn* 4, 58–64, *SC* 156, p. 194. *The Ethical Discourses* vol. 13, 131–133, *SC* 129, p. 410. *Hymn 53*, 112–132, *SC* 196, p. 220.

96. Symeon the New Theologian, *The Ethical Discourses* vol. 10, 372–374, *SC* 129, p. 286.

97. Symeon the New Theologian, *The Secret Wish*, *SC* 156, p. 40. *Catechesis* 16, 91–92, *SC* 104, p. 246. *The Ethical Discourses* vol. 5, 350–353, *SC* 129, pp. 104–106.

98. Gregory the Theologian, *On Holy Easter*, *PG* 36, 632AB.

99. Gregory Palamas, *Homily* 60,2, ed. S. Oikonomou (Athens: 1861), p. 248, 249. Symeon the New Theologian, *The Ethical Discourses* vol. 6, 161–178, *SC* 129, pp. 130–132.

100. Symeon the New Theologian, *The Ethical Discourses* vol. 10, 769–773, *SC* 129, p. 314. Symeon the New Theologian, *The Ethical Discourses* vol. 4, 665–668, *SC* 129, p. 56.

101. Symeon the New Theologian, *The Discourses* vol. 2, 209–218, *SC* 122, pp. 144–146. *The Ethical Discourses* vol. 4, 673–677, *SC* 129, p. 56. Symeon the New Theologian, *The Discourses* vol. 1, 217–222, *SC* 122, pp. 110–112. Epistle to Timothy 1,2,9. Isaiah 61,3.

102. Gregory Palamas, *Epistle to Akindynos* 3,16, ed. P. Christou. Gregory Palamas, *Writings*, vol. 1 (Thessaloniki: 1962), p. 308. Matthew 5,14.

103. Dionysius the Areopagite, *On the Ecclesiastical Hierarchy* 4, *PG* 3, 376A.

104. Maximus the Confessor, *Various Chapters* 1,41, *PG* 90, 1193D.

105. Maximus the Confessor, *Various Chapters* 1,75, *PG* 90, 1209C.

106. Symeon the New Theologian, *Catechesis* 15, 68–99, *SC* 104, p. 228.

107. Symeon the New Theologian, *The Ethical Discourses* vol. 6, 137–138, *SC* 129, p. 130. *Hymn* 52, 1–6, *SC* 196, p. 198.

108. Symeon the New Theologian, *The Ethical Discourses* vol. 8, 90–98, *SC* 129, p. 208.

109. Vladimir Lossky, *The Vision of God*, trans. M. Kalamaras (Rigopoulos, Thessaloniki: 1973), p. 189. Symeon the New Theologian, *The Ethical Discourses* vol. 1, 61–68, *SC* 122, p. 228.

110. Gregory of Nazianzus, *Epitaph to Caesarion, Discourse* 7,23, *PG* 35, 785.

111. Symeon the New Theologian, *The Discourses vol. 1*, 360, *SC* 122, p. 122. *Hymn 23*, 26–38, *SC* 174, p. 190.

112. Anthony Cramer, *Catenae Graecorum Patrum in Novum Testamentum*, ed. University of Michigan Library, 2009, p. 270. Konstantinos Skouteris, *Man According to the Word of the Divine Economy* [http://www.apostoliki-diakonia.gr/bible/bible.asp?contents=themata/contents.asp&main=diafora&file=2.2.htm (08/01/2021)].

113. Gregorius Nyssenus, *Theol. De Opificio Hominis,* (2017.079) (A.D.4), p. 185, 32.

114. Second Epistle to the Corinthians 5,17.

115. John 5,24. John Chrysostom, *Homily on Romans*, 10,1, *PG* 60,174. Colossians 3,2–3.

116. Konstantinos Skouteris, *Man According to the Word of the Divine Economy.*

117. Gregorius Nazianzenus, Theol. *In Sanctum Pascha (orat. 45)* (2022.052), volume 36, p. 633,51.

118. Colossians 3,9–10.

119. Ephesians 4,23.

120. Basil the Great, *On the Holy Spirit*, 15,35, *PG* 32,128.

121. Symeon the New Theologian, *Ethical Discourses*, 6, 177–178, *SC* 129, p. 120.

122. Athanasius the Great, *Against the Arians* 1, 16, *PG* 26, 45BC.

123. John Chrysostom, *To the Scandalised* 1, *PG* 94, 984.

124. John 11,25–26. Ephesians 2,4–7.

125. George Mantzaridis George, *Christian Ethics II*, p. 50.

126. Gregory Palamas, *Homily*, 52,2, Oikonomou (Athens: 1861), p. 122.

127. Basil the Great, *On the Holy Spirit*, 15,35, *PG* 32, 128CD.

128. Symeon the New Theologian, *Ethical Discourses* 5, 33–34, *SC* 129, p. 82. *Catechesis* 5, 395–397, *SC* 96, pp. 401–406, 410. *Hymn* 42, 188–192, *SC* 196, p. 52.

129. Cyril of Alexandria, *On the Holy Trinity*, 3, *PG* 75, 808C.

130. John 3,6.

131. Gregory Palamas, *In Favour of the Holy Resting* 1,3,43, ed. Christou P., Gregory Palamas, *Writings*, vol. 1 (Thessaloniki: 1962), p. 132.

132. Irenaeus of Lyons, *Control*, 5,6,1, *PG* 7, 1137A.

133. George Mantzaridis, *Christian Ethics II*, p. 676.

134. Symeon the New Theologian, *Hymn* 50, 203–212, *SC* 196, pp. 170–172. *Hymn* 26, 42–50, *SC* 174, p. 272.

135. Symeon the New Theologian, *The Ethical Discourses,* vol. 2, 218–219, *SC* 122, p. 382.

136. George Mantzaridis, *Christian Ethics II*, p. 511.

137. Symeon the New Theologian, *Hymn* 46, 27–37, *SC* 196, p. 116.

138. Nicholas Kabasilas, *On the Life in Christ*, 6, *PG* 150, 680C. George Mantzaridis, *Christian Ethics II*, p. 677.

139. Skouteris, *Man According to the Word of the Divine Economy.*

140. First Epistle to the Thessalonians 5,23.

141. First Epistle to the Corinthians 6,19.

142. Ephesians 4,22.

143. Colossians 3,5.

144. Colossians 3,9–10.

145. Colossians 3,11. Ephesians 2,15.

146. Ephesians 4,13.

147. Titus 3,5.

148. Galatians 3,27. Skouteris, *Man According to the Word of the Divine Economy*. Epistle to the Romans 6,3–6.

149. Second Epistle of Peter 1,4. John 14,6. Ioannis Karmiris, *Orthodox Ecclesiology, Dogmatics Part E'* (Athens: 1973), p. 141.

150. Skouteris, *Man According to the Word of the Divine Economy*.

# Chapter 11

# Ecclesiology

## *The Nature of the Orthodox Church*

According to the Encyclical of the Holy and Great Council of the Orthodox Church (Crete 2016), the One, Holy, Catholic, and Apostolic Church as the body of Christ in the image of the Holy Trinity, is the God-human communion of the Holy Trinity, experiencing, foretasting of the Last Days through the Eucharist, and revealing the glory of those to come. It is a perpetual Pentecost and an unswerving prophetic voice; it is the testimony and presence of "the Kingdom of God's revelation in power."[1] The Church, as the body of Christ, "attaches"[2] to Him, infuses and transforms the world.[3]

The Orthodox Church is true to the unanimous Apostolic Tradition and sacramental experience, and is the genuine continuation of the One, Holy, Catholic and Apostolic Church, as it is confessed in the Creed and affirmed by the teaching of the Fathers. The Church preserves the authentic lived experience in Christ and the credible testimony of the truth to all people within the ecclesiastical body.[4]

The Church is constituted as a Synod under Christ and guided by the Holy Spirit.[5] The Church, through the Ecumenical and Local Councils, proclaims the mystery of the Holy Trinity, which was revealed through the incarnation of the Son and Word of God.[6] The holiness of a human takes place within the body of Christ, which is the Church.[7] Holiness comes from the Holy God; it is the participation of a human in the holiness of God "in the communion of holies."[8]

The Church preexisted, from time immemorial, in the will and wisdom of God under His "counsel" and "providence," before the foundation of the world, and at creation it included the mental spiritual world, the angels, and then humans, and relates to the age-old divine will for the salvation of the world.[9]

The Church has a Trinitarian character, as revealed in the New Testament, being the people of God the Father, the body of Christ and the temple and communion of the Holy Spirit. The triune nature of the Church is confirmed by the character of the initiatory sacrament, since we are baptized "in the name of the Father, the Son and the Holy Spirit" and the baptized person, by the grace of the Father, puts on Christ through the Holy Spirit. Thus, just as the Church is the dwelling place of the Holy Trinity, so too the believer, as a member of the body of the Church, becomes the indwelling place of the Holy Trinity.[10]

Everything in the Church comes from the Father through the Son and in the Holy Spirit. At the foundation of the Church lies God the Father, as the cause and beginning of everything, the Son through whom all things were made, and the Church as His body, and the Holy Spirit as the soul, the life-giving and sanctifying power which acts on everything in the Church.[11]

In the setting of the Divine Economy, where one person of the Holy Trinity acts, the other persons are also present, indicating its unity. Consequently, "where there are three, the Father, the Son, and the Holy Spirit, there is the Church, which is the body of all three,"[12] and "wherever one of the Trinity is present, all of the Trinity is present, for it is undivided from itself and united with precision."[13]

The Church preexists eternally in the immemorial will of the Father, and in His immemorial will for the creation of the spiritual and physical world, the creative power of the Son and Word of God is also included, in the context of the cooperation of the members of the Holy Trinity. The founder of the Church, the Son, and Word of God, preexisted eternally as Christ in the immemorial will of God the Father, prior to the creation of the world and human, as the firstborn in the Church of the firstborns.[14]

The Church is an image of the Holy Trinity which is communicated as the coming of the Trinitarian Economy,[15] as a revelation from the Father through the Son in the Holy Spirit. God reveals Himself as a Monad in essence and as a Trinity in substance, as a supreme unity of Persons communicating inner love and unity. This act of intratrinitarian love is manifested and communicated to the outside world in the Church as a similitude of the perfect intratrinitarian love. The Church communicates divine love which is the unifying factor of the multiple persons. In this sense, the Church is a place of unity and love, as a form and image of God.[16]

In the setting of the Divine Economy, the Church of the Old Testament was the "shadow of the good things to come"[17] and in the New Testament period, the fulfilment of the promise and the redemptive expectation of the whole world took place. According to the apostolic and patristic tradition, the Church is the eternal historical body of Christ, which the Son and Word of God received at his incarnation "under pure conception" from the Holy Spirit through the Virgin Mary. Christ, being incarnate, assumed the flesh of the Church.[18]

Christ is the head of the Church and "all things are created through Him and for Him, and He is before all things and in him all things hold together, and He is the head of the body of the Church."[19] It is upon the Church that the unity of creation is built and serves as the cohesive force of the world, and through His economy the God-Human paves the way for the reunification and reconstruction of creation.[20]

Through the incarnation of the Son of God, the "Church took on flesh," built up through His incarnation and Resurrection. The Church is the body of the incarnate God who suffered, was resurrected, and ascended in glory into the heavens. Within the Church the human has been united with the divine, the invisible with the visible, the transitory and temporal with the eternal. God was united with humans in a genuine, mystical unity, in a living mystical body, the Church, whose members are all the Christians, and Christ the head of all.[21]

Therefore, within the Church, the unifying power of the Divine energy operates which brings together all sensible and mental beings in a unity which is dependent on the unity by grace with the uncreated God.[22]

Maximus in his ecclesiology approaches the Church with a cosmological iconography, consisting of invisible and visible entities as an image of the universal world. The world of beings, created by God, consists of the mental world, which is composed of mental and incorporeal forces, and of the corporeal or sensible world.[23] Since the Church is a form and image of the Holy God and God is the origin of the universal world, it functions as the cohesive and sanctifying power of the universal world.[24]

The Church includes the entire invisible and visible creation, which constitutes and permeates the whole world, both mental and tangible, as well as human in body and soul. According to Maximus the Confessor, the Church is a symbol, form, and image of the Triune God, of the visible and invisible world, of humans in general, a form and image of the soul, a charismatic communion of humans and the world.[25]

The Church lives and exists both on earth and in heaven, for one of its aspects is heavenly and invisible and the other is visible, earthly, and historical communion. It is the on earth devoted and heavenly triumphant Church as a communion of Saints where the earthly members come to the heavenly Jerusalem.[26]

According to Professor Vlasios Feidas, the Ecclesiology of the Eastern Orthodox Church has been shaped based on the biblical, apostolic, liturgical, and patristic tradition and in accordance with the decisions of the Ecumenical Councils. Moreover, "Christology is the ultimate, unchanging and immutable solid theological core, on the one hand for Trinitarianism and Pneumatology, and on the other hand for Ecclesiology and anthropology."[27]

## THE CHURCH AND HUMANITY

Within the Church the mystery of the salvation of humanity and of all the creation is revealed.[28] The world was created, and humanity was formed by God's will, to participate in the coming of the divine life, so that the union of human will with God's will may be realized.[29]

It is understood that following the Fall, humans and creation find their closure and the potential for completion and fulfillment of their destiny within the charismatic Body that is the Church of Christ. In the unity of the body of Christ, the Church, grace is provided through the Holy Spirit to both humans and creation, so that they can commune and unite with God, along with the eschatological prospect of the heavenly Kingdom of God.[30]

The Church is a workshop and ark of salvation, where every person can attain it through the freely exercised faith in love, but also through the participation of the Holy Sacraments and the perfecting and fruitful work of Divine Grace towards salvation.[31] Within the Church, the redemptive act is carried out, with Christ and His prophetic, high-priestly, and royal energy at its center, and with the vital force of divine grace and truth in the Holy Spirit, the mission of the Church, as the ark of salvation, is realized and brought to fruition.[32]

The Church governs everything, is present in all creative manifestations, and acts in a sanctifying and unifying way through uncreated actions so that, within it, the communion of creation with God and the deification of man, which ensures sanctification and salvation, takes place.[33] In this way, a person can become a "temple of God" through the Holy Spirit to radiate the glory of God.[34] According to Nicodemus the Hagiorite, the soteriological and sanctifying dimension of the Church has theoretical and practical references. The Church is the home of all Christians, the house of God, the place where the peace of Christ reigns, the Jerusalem, and the new Zion. The devoted Church is the image and form of the triumphant Church.[35]

The Church is a living divine-human organism and body of Christ, in which Christ and the Christians throughout the ages are organically united, just as the human and the divine nature were united in the person of Christ. Thus, all Christians are united in a genuine and mystical unity with the head, Christ, in an organic unity and wholeness in a mystical body.[36]

Within the Church the restoration of human's communion with God occurs, the mystical-sacramental realization of an arc of love, the manifestation of the new creation, and a restorative regeneration of the whole world.[37] In the Church, the divine love is experienced, and the believers commune with the one and indivisible Christ.[38]

The Church is the form and image of God because, within it, the unification of believers into a single body is realized, despite their differences, in the way that in "God's nature," He unites and harmonizes everything as the origin, beginning, and end.

The Church is an image of the Triune God, within which the similitude with God and the union of everything is realized. Within the Church the believer becomes God by grace, that is, he is asked to be deified devoid of the identity of his essence with God. Thus, theosis-deification has an ecclesiological character, being considered as "communion of theosis."[39]

The Church is One and indivisible, which unites its members through the bond of love and the truth of the Holy Spirit.[40] The members of the Church are united to Christ and are called upon to unite with one another according to the standard of the Holy Trinity.[41]

The Church is One and indivisible and consists of the spiritual, divine, and invisible elements and the visible, material, and human elements. The unity exists in the sense of the mystical and distinct union of human nature with the divine nature in the person of Christ. In the unseen part of the head lies Christ, as the Holy Spirit dwells and acts with its sanctifying and salvific energy, and in the visible part lie the believers, the devoted body of the Church, to which grace is imparted to ensure the salvation of its members.[42]

The likeness with God, based on the ecclesiological approach, apart from the union of the members, also involves the inner union of the person torn by passions and sin. The functions of the soul and the body become divinized following the paradigm of the unity of the two natures of Jesus Christ. Consequently, unity within the Church refers to the unity which results from the communion of a person with God, the unity of communion of humans, and the unity of a human himself with the bond of godlike love, which leads to the unity of the universal world within the Church.

According to Saint Maximus, within the Church the believers are united both with each other and with God. The believers are connected to one another crosswise, that is, initially in a horizontal way, as well as in a vertical way, as a union between people, of the visible and invisible world with one another and with God. In this context, the cohesive and unifying Divine action of the Church between the mental and the sensible is revealed as the communion of the invisible and visible world, just as the dual human being with body and soul is and can partake of the Divine actions. Moreover, in this unity Christ, through the Divine actions, brings all beings into a harmony with the uncreated God and eliminates all distinctions between created and uncreated nature.[43]

In keeping with the unity of the members of the Holy Trinity, the unity of the believers within the Church is also achieved as a sanctifying, transforming, unifying, and loving communion of people in Christ through the

Holy Spirit with the benevolence of the Father.[44] Along with the unity of the believers within the Church, as members of the body of Christ, each person's identity, uniqueness, and autonomy are preserved in line with their characteristics.[45]

Additionally, within the Church, aside from the communion of humans with God, the communion of individuals with each other is ensured and realized. Humanity steps into a communion of unity and love with his fellow humans, viewing a person as an image of God.[46] Love at this level is considered as a depiction of God's attributes or actions, but also as a structural characteristic of humans, who, when he practices it, finds his true self according to God and becomes akin to Him within the body of the Church following the standard of the unity of the Holy Trinity.

Within the Church, humanity experiences unity by having a conscience and a sense of communion with God. Having faith as a pillar, since "faith is a transcendental relationship, through which we are unknowingly, but not demonstrably, united with God by reason" and through the cultivation of virtues and participation in the Sacraments, the believer partakes in the actions within the Church. Church, as the image of God and the body of Christ, functions unifyingly and cohesively, bridging the gap caused by sin between humans and God, eternity and time, and fosters human relationships in communion with each other and God the creator.[47]

The Church is a functional unity of the world and humanity; it governs everything, and through its uncreated actions, it serves a sanctifying and unifying role as the communion of persons and creation with God. As a communion, it governs all creative manifestations and channels its sanctifying energy due to the holiness of its Head, which is Christ. As the treasure of wisdom and knowledge, it opens the mind of a human "to the knowledge of the mystery of God and Christ's Father, in whom all the treasures of wisdom and knowledge are hidden."[48]

Just as every person of the Holy Trinity is a perfect God, so does every believer as a member of the Church contain all humanity in himself and become an ideal human by participating in holiness.[49]

The "supernatural true beauty of the Church"[50] is the glory of the Triune God, which is imparted to the cleansed through Baptism "in the image" human, as the likeness of God in the Holy Spirit. The Holy Spirit bestows incorruptibility and holiness as an attribute of God, which humans commune it as his image, and based on the likeness, he is nurtured towards perfection.[51]

Spiritual life is cultivated within the Church, in which every believer is called upon to become a communicant of the Triune God in Christ through the grace of the Holy Spirit. It is within the Church that the communion of humans, as created in the image of God, is established and comes to fruition. The Trinitarian model applies collectively to the Church and every believer because the renewal in Christ takes place through the Holy Spirit. The life of the Church is charismatic with an eschatological perspective.[52] Within the Church, humanity is renewed and transformed by God the Father's providential, salvific, and sanctifying action through Jesus Christ in the Holy Spirit.

Of the creations, some share in the essence-giving energy of God, as they simply exist as beings; others share in the life-giving energy, for they exist as living beings, and others share in the wisdom-giving energy as rational beings, while angels and saints share in the God-giving Divine energy.[53]

The believer within the Church participates in the benevolence and truth in the likeness of God. Benevolence pertains to the practical power of the soul, while truth to the theoretical. Thus, when the soul, and consequently the whole person, take the rational action as "theory acted upon," in combination with the practical knowledge as a "sacramental act," he is led to spiritual perfection within the body of the Church.[54]

The sanctification and theosis-deification of humans are accomplished within the Church, as the Church of Christ is "all things and in all things."[55] Since Christ is the vine and the believers are the branches, they bear fruit when they stay united with Christ[56] and are led beyond fruitfulness to salvation. The Church is organically connected to Christology since Christ is bound to the indivisible body of the One, Holy, Catholic, and Apostolic Church. The Christocentric character of the Church is also defined by the direct connection of the Church with the Sacrament of Eucharist and by the entire sacramental life with the soteriological prospect for each member of the Church.

## THE DEVOTIONAL AND SACRAMENTAL ASPECTS OF THE CHURCH

The Church, as the body of Christ, is a mystical and divinely inspired organism whose liturgical life is devotional and sacramental. The devotional life culminates before the holy altar, at the holy table, in the holy temple, where all the believers are gathered and summoned to the services and communal worship.[57]

The Church also means the assembly of people, the gathering of believers in the temple of God.[58] The place of assembly is, in a way, the ark where those who come are saved from the spiritual deluge of sins and passions.[59]

When Christians enter the house of God, as the visible Church, it is as if they were entering heaven, of which the holy temple is an image and form. The believers stand before God, asking for mercy and forgiveness for their sins, united in only the body of Christ, the Church.[60]

The Church is the infirmary that heals sinners from spiritual wounds. It provides spiritual medicine and food through the Sacraments and its teaching. Consequently, it is the ark that provides salvation, and it is the spiritual manger in which the flock of God is protected, healed, and regenerated.[61]

Within the Church, the communication with God and the communion of the risen Christ's body and blood take place, so that the believers may maintain the purity of the soul[62] and receive spiritual strength.[63]

The Church is the place of prayer, the house of God, the place of spiritual rebirth, the place where the Holy Sacraments are performed, the place of worship and praise of God, and the veneration of the Saints. It is the place and the vehicle by which Christians aim to receive God's grace; it is the center of unity, prayer, and the practice of spiritual life.[64]

Within the Church, the Holy Sacraments are performed for all believers, the Doctrines of Faith are preached, and the Divine Commandments, the Scriptures, and the Word of God are interpreted. In addition to spiritual healing, the believer receives spiritual strength, the soul is purified, the soul and body are sanctified, and the entire human.[65]

The physical and spiritual space of the Church is a place of unity where Christians, through prayer and the Sacraments, as well as the practice of the virtues, are united in the one body of Christ's Church to the fellow human and God.[66]

According to Chrysostom, the Church is not only a place but also a way, beyond the walls it has laws. When one takes refuge in the Church, they experience that it is not a mere place of refuge but a place of conviction and faith that defines and determines life.[67] He also urges believers not to abstain from the Church since the Church is hope, refuge, and salvation. There is nothing mightier, for it is higher than heaven and broader than the earth, never growing old and always flourishing. According to the Scriptures, the Church is described by many names symbolizing its incorporeal qualities, as the unmovable and solid mountain, the immaculate, the virgin, the queen, the sterile one who begets, the fountain, the multiparous, the lily, as well as by many other names to express its holiness and nobility.[68] The salvific and sanctifying work of the Church, apart from its spiritual teaching, is also accomplished through its Sacraments.

The Sacraments are the created means, the "doorways" through which the light of God enters the world, and the uncreated grace of God is provided. The Sacraments of Baptism, Chrismation, and the Eucharist recapitulate the Divine Economy in Christ.[69] The fruition of the Holy Sacraments requires the synergy and consent of the believer, for the divine Grace to work and lead the believer to the completion of life in Christ.

Christ shows mercy to people in the Church through its Sacraments, with the forgiveness of sins, the release from death and corruption, granting them salvation.[70] Cyprian of Carthage points out that outside Church there is no salvation.[71]

Within the Church's sacramental life, a person's transformation is evoked, and corruption is overcome. The Sacraments, with their outward signs, are organically connected with the Archetype and the spiritual dimension of the celebrants as a mystical union of the visible with the invisible, the tangible with the mental.[72] A human, as a created being by the grace of the Sacraments of the Church, can become like God by grace through the uncreated Divine actions.[73] Through Baptism, Chrismation, and the Eucharist, humans become partakers of the gifts of the Holy Spirit and members of the body of Christ.

Through the baptism of Christ, humans are regenerated, reborn, and clothed with divine glory. Through repentance and spiritual life, the grace of the Holy Spirit is provided, enabling the believer to be ready to be united in the Holy Spirit with Christ in the Church through the Sacrament of Holy Communion.

Within the body of the Church, with Christ as its head, the baptized are incorporated "in one spirit"[74] and through the believers' participation in the Sacrament of the Eucharist, as well as the other sacraments, they become vital members of the body of the Church, in which Christ is the head and the savior of the body.

The "life in Christ" is directly related to the holy Sacraments of Baptism, Holy Communion, and the rest of the sacraments since humans participate in the body of the Church by the grace of the Holy Spirit.[75]

The communion of persons is realized within the Church in Christ. A true communion is created because the individuals partake of Christ, are incorporated into His body, remain united with Him, and are enlivened by His presence.[76] The life and will of the believer are guided by Christ and nourished by His body and blood.[77]

Spiritual fulfillment occurs within the Church, where the believer, mentally and physically, partakes of the body of Christ through the Sacraments and is nourished, assembled, and renewed mystically.[78] During the Communion of the life-giving and immaculate sacraments, the unity is realized, whereby a human, because of his virtual "in the image" relationship with God and the possibility of resembling Him, is honored by becoming a "god" by grace instead of a person.[79] The bread and wine of the Eucharist, the Eucharistic body of Christ, is the risen and eternal historical body of Christ assumed into heaven.[80]

According to John Chrysostom, the honorable gifts of the Eucharist are brought to the heavenly altar to be united with the ascended eternal historical body of Christ "that is the Church."[81] The honorable gifts are consecrated by the invocation of the Holy Spirit, which asserts the synergy of the Holy Spirit in the Eucharistic body of Christ and works towards the salvation of humans.

Besides their faith in Christ, Christians are incorporated into His body initially through Baptism and Chrismation, spiritually nourished by the Eucharist, and sanctified through the other sacraments, as the grace of the Holy Spirit brings about an inner and perfect organic unity among them.[82] Through the Eucharist, the members of the Church become partakers of the body and blood of Christ, and they become "whole" and members of his body, since "to Christ one flesh through the Eucharist."[83]

Through the Eucharist, the Church is revealed as the body of Christ and the communion of the Holy Spirit, in which the earthly and the heavenly Churches are united, the "present" is united with the "future era" as an ecclesiastical eschatological experience that lives the eschata in the present.[84]

Therefore, along with Christocentrism, the energy and presence of the Holy Spirit in the Church is also essential because the Paraclete sanctifies and graces the Church, as this was particularly manifested at Pentecost when the gifts of the Holy Spirit were poured out, and continues to remain in the Church and to sanctify its members unceasingly.[85] The Sacraments of the Church are administered by the power, the revelation, and the energy of the Holy Spirit for the sanctification and salvation of the believers, the members of the Church, and the members of the body of Christ.[86]

In the unity of the body of Christ, the unity of humanity was recapitulated, where the believers, through their communion in the Eucharist and through their association with this "mystery of the sacraments," participation in the other sacraments, within the sacramental life of the Church experienced as the body of Christ, are inspired by the hope of the Resurrection and by the expectation of His glorious Second Coming. The unity of the body of Christ within the Church through the sacramental life has an ecclesiological and soteriological background since all of humanity is united both spiritually and visibly until the end of the ages through the synergy of the Son and the Holy Spirit and the favor of the Father.[87]

The Grace, sanctification, and salvation of the members of the Church are communicated and granted by God and the Father through Christ in the Holy Spirit, as "the Father gives all things through the Son in the Spirit"[88] and "the Father and the Son and the Holy Spirit have administered the economy intended for us."[89]

Throughout the worship of the Church, the Holy Trinity is praised, "God is glorified through Christ in the Holy Spirit"[90] in a multitude of Trinitarian hymns and the Doxology, while the commencement of the liturgical celebration of the Sacraments and their subsequent hierology have a clear Trinitarian character.

## NOTES

1. Official Documents of the Holy and Great Council of the Orthodox Church, *Encyclical I. The Church: The Body of Christ, in the Image of the Holy Trinity* §1, Crete, June 2016 [https://www.holycouncil.org/-/encyclical-holy-council (19-12-2020)]. Mark 9,1.
2. Matthew 23,37.
3. Official Documents of the Holy and Great Council of the Orthodox Church, *Encyclical I. The Church: The Body of Christ, in the Image of the Holy Trinity* § 1.
4. Official Documents of the Holy and Great Council of the Orthodox Church, *Encyclical I. The Church: The Body of Christ, in the Image of the Holy Trinity* § 2.
5. Acts 15,28.
6. Official Documents of the Holy and Great Council of the Orthodox Church, *Encyclical I. The Church: The Body of Christ, in the Image of the Holy Trinity* § 3.
7. Ephesians 1,23.
8. Official Documents of the Holy and Great Council of the Orthodox Church, *Encyclical I. The Church: The Body of Christ, in the Image of the Holy Trinity* § 4.
9. Ioannis Karmiris, *Orthodox Ecclesiology, Dogmatics Part E'*, Athens 1973, pp. 22, 23. Clement of Alexandria, *Paedagogus* 1,6, *PG* 8, 21.
10. John Chrysostom, *On the Gospel of John* 25,2, *PG* 59, 159. Ioannis Karmiris, *Orthodox Ecclesiology, Dogmatics Part E'* (Athens: 1973), p. 200.
11. Ioannis Karmiris, *Orthodox Ecclesiology, Dogmatics Part E'*, p. 201.
12. Tertullian, *De Baptism 6*, Corpus Scriptorum Ecclesiasticorum Latinorum (CSEL), editume consilio et empensis Academiae Litterarum Caesareae (Vindobonensis: 1866), CSEL 20, 206.
13. John Chrysostom, *Homily on the Holy Pentecost*, 1,2, *PG* 50, 456.
14. Vlasios Feidas, *Ecclesiology between Christology and Pneumatology in the Light of the Patristic Tradition* (Athens: 2018) pp. 19–21. Ephesians 3,3–10.
15. Ephesians 1,23.
16. F. Dumitru Stăniloae, *God, the World and Man*, ed. Armos (Athens: 1990), p. 119.
17. Hebrews 10,1.
18. Vl. Feidas, *Ecclesiology between Christology and Pneumatology in the Light of the Patristic Tradition*, pp. 22, 23, 63. John Chrysostom, *Before the Exile*, *PG* 52, 35.
19. Colossians 1,16–18.
20. Konstantinos Skouteris, *Man According to the Word of the Divine Economy*.
21. Ioannis Karmiris, *The Ecclesiology of the Three Hierarchs* (Athens: 1962), p. 82. Maximus the Confessor, *Chapters on Theology*, *PG* 90, 1136.
22. Maximus the Confessor, *Mystagogy*, pp. 106, 119. Maximus the Confessor, *To Thalassios on Difficulties in Sacred Scripture*, *PG* 90, 436.
23. Maximus the Confessor, *Mystagogy*, pp. 117, 118.
24. Maximus the Confessor, *Mystagogy*, p. 107.
25. Maximus the Confessor, *Mystagogy*, pp. 106–126.
26. Clemenes of Alexandria, *Paedagogus* II, 1, *PG* 8, 388. Ioannis Karmiris, *Synopsis of the Doctrine of the Orthodox Catholic Church*, pp. 78, 19.

27. Vl. Feidas, *Ecclesiology between Christology and Pneumatology in the Light of the Patristic Tradition*, pp. 22, 60.

28. Vladimir Lossky, *The Mystical Theology of the Eastern Church*, p. 118.

29. Aristarchos Grekas Archimandrite, *The Church as a Place of Communion between God, Man and the World*, p. 30.

30. Vladimir Lossky, *The Vision of God*, Holy Metropolis of Nicopolis (Preveza: 2004), p. 127.

31. Ioannis Karmiris, *Synopsis of the Doctrine of the Orthodox Catholic Church*, p. 51. John Chrysostom, *On Genesis, Homily*, 16, 5.6, *PG* 53, 131.

32. Ioannis Karmiris, *Synopsis of the Doctrine of the Orthodox Catholic Church*, p. 83.

33. Colossians 2,3. Aristarchos Grekas Archimandrite, *The Church as a Place of Communion between God, Man and the World*, p. 30.

34. First Epistle to the Corinthians 3,18.

35. Nicodemos the Hagiorite, *Eortodromion (or Commentary on the Canons of the Dominical and Marian Feasts)*, vol. B, Orthodoxos Kypseli (Thessaloniki: 1995), pp. 29, 188. Nicodemos the Hagiorite, *New Klimax (or Commentary on the Seventy-FIve Anavathmoi of the Octoechos),* Schinas (Volos: 1956), pp. 62, 159. Nicodemos the Hagiorite, *Ethics of Christians*, Rigopoulos (Thessaloniki: 1998), pp. 159, 296, 302. Dionysius the Areopagite, *Epistle 10*, *PG* 3, 1117A–1117B. First Epistle to Timothy 3,15. Hebrews 12,22.

36. Ioannis Karmiris, *Synopsis of the Doctrine of the Orthodox Catholic Church*, p. 79. Colossians 2,19.

37. Second Epistle to the Corinthians 5,17.

38. Symeon the New Theologian, *Chapters* 3,3, *SC* 51 (Paris: 1980), p. 120.

39. Gregory Palamas, *Demonstrative Discources on the Procession of the Holy Spirit*, vol. 2, σ. 149. Maximus the Confessor, *To Thalassios*, 61, *PG* 90, 44D. Maximus the Confessor approaches the Church as an image of man in which the body is the temple, the mind is the holy altar, and the soul is the sacred step, as well as the Church as an image and form of the soul. Maximus the Confessor, *Mystagogy*, pp. 126, 156–158.

40. George Mantzaridis, *Christian Ethics II*, p. 328.

41. Gregory Palamas, *Demonstrative Discources on the Procession of the Holy Spirit*, vol. 2, p. 149.

42. Ioannis Karmiris, *Synopsis of the Doctrine of the Orthodox Catholic Church*, p. 81.

43. Maximus the Confessor, *To Thalassios*, 48, *PG* 90, 436. Aristarchos Gkrekas Archimandrite, *The Church as a Place of Communion between God, Man and the World*, p. 18.

44. Cyril of Alexandria, *Commentary on John*, 17, 20–21, *PG* 74, 557D. *Commentary on Peter*, 2, 6, 7, *PG* 74, 1013A. Nikolaos Matsoukas, *Dogmatic and Symbolic Theology B'. Exposition of the Orthodox Faith in Contrast with Western Christianity, Philosophical and Theological Library 3* (Pournaras, Thessaloniki: 2001), p. 137.

45. Cyril of Alexandria, *Commentary on John*, *PG* 73, 65A. Father Dumitru Stăniloae, *The Orthodox Christian Teaching on God*, ed. Armos (Athens: 2011), p.

47. Ioannis Panagopoulos, "Ontology or Theology of the Person. The Contribution of Patristic Trinitarianism to the Understanding of the Humans," *Synaxis*, 14 (1985): 35–47. Chrysostomos Stamoulis, *About Light. Personal or Physical Energies? Contribution to the Modern Problematic Holy Trinity in the Orthodox Space*, To Palimpsiston (Thessaloniki: 1999), pp. 65–86.

46. Skouteris, *Man According to the Word of the Divine Economy.*

47. Skouteris, *Man According to the Word of the Divine Economy.*

48. Colossians 2,2–3.

49. George Mantzaridis, *Christian Ethics II, Man and God, Man and Fellow Man, Existential and Bioethic Propositions and Perspectives*, p. 139.

50. Cyril of Alexandria, *On the Veneration and Worship in Spirit and Truth*, B', *PG* 68, 238D.

51. Cyril of Alexandria, *On the Veneration and Worship in Spirit and Truth,* A', *PG* 68, 145D, 149B. *Commentary on Isaiah*, A, 6, *PG* 70, 236B. *On the Holy and Consubstantial Trinity*, A, 3, *PG* 75, 853C.

52. George Mantzaridis, *Christian Ethics II, Man and God, Man and Fellow Man, Existential and Bioethic Propositions and Perspectives*, pp. 301, 316. Luke 17,21.

53. Gregory Palamas, *Disputation to Akindynos*, vol. 3, P. Christou (Thessaloniki: 1970), p. 375.

54. Nikolaos Matsoukas, *Dogmatic and Symbolic Theology A'. Introduction to Theological Epistemology* (Pournara, Thessaloniki: 1985), p. 215. Maximus the Confessor, *Mystagogy*, p. 185. *Various Chapters on Theology and Economy*, *PG* 90, 1344.

55. Colossians 3,11.

56. John 15,1–2.

57. Nicodemos the Hagiorite, *New Klimax*, p. 135.

58. Nicodemos the Hagiorite, *Ethics of Christians*, pp. 300, 301, 388. St. Germanus, *Ecclesiastical History and Secret Theory* 1, *PG* 98, 384B.

59. Nicodemus the Hagiorite, *Ethics of Christians*, pp. 294, 307. *New Klimax*, p. 298. John Chrysostom, *On the Earthquake and the Rich Man and Lazarus* 7, *PG* 48, 1037–1038.

60. Nicodemos the Hagiorite, *Ethics of Christians*, pp. 176, 298, 308, 346, 397. Symeon of Thessalonica, *On the Holy Temple*, CXXXVI, *PG* 155, 345C–345D. Nicodemus the Hagiorite, *The Rudder (Pedalion) of the Metaphorical Ship of the One, Holy, Catholic and Apostolic Church of the Orthodox Christians*, 8th reprint of the 3rd edition of 1846, Rigopoulos (Athens: 1976), pp. 328–330.

61. Nicodemus the Hagiorite, *Ethics of Christians*, pp. 278–281. John Chrysostom, *On St. John*, 2,5, *PG* 59, 36. *On Genesis* 32,1, *PG* 53, 293.

62. Nicodemos the Hagiorite, *Ethics of Christians*, pp. 326, 327. *New Klimax*, pp. 212, 213.

63. Nicodemos the Hagiorite, *New Klimax*, p. 446. Maximus the Confessor, *Mystagogy*, 24, *PG* 91, 701D–704A.

64. Nicodemos the Hagiorite, *Ethics of Christians,* p. 33.

65. Nicodemos the Hagiorite, *Ethics of Christians,* pp. 326, 327. *New Klimax*, pp. 212, 213. *Garden of Graces (or an Elegant Interpretation of the Nine Odes of the*

*Stichologia)*, Rigopoulos (Athens: 1992), p. 154. Maximus the Confessor, *Mystagogy*, 24, *PG* 91, 701D–704A.

66. Nicodemos the Hagiorite, *Ethics of Christians*, p. 346.

67. John Chrysostom, *Homily. After Eutropius Having Been Found Outside the Church Had Been Taken Captive, and on Paradise and the Scriptures*, *PG* 52, 397.

68. John Chrysostom, *Homily. After Eutropius Having Been Found Outside the Church Had Been Taken Captive, and on Paradise and the Scriptures*, *PG* 52, 402, 403.

69. Nicholas Kabasilas, *On the Life in Christ*, 1, *PG* 150, 504AB.

70. Vladimir Lossky, *The Mystical Theology of the Eastern Church*, p. 217. Symeon the New Theologian, *The Ethical Discourses* vol. 13, 170–171, *SC* 129, p. 412.

71. Cyprian of Carthage, *Epistle*, 73,21, *PL* 3, 1169A.

72. Maximus the Confessor, *Mystagogy*, p. 224.

73. Verna Harrison, "Receptacle Imagery in St. Gregory of Nyssa's Anthropology," *SP 22* (Leuven: 1989), p. 24. Gregory of Nyssa, *On the Soul and the Resurrection*, *PG* 46, 105A. Gregoire De Nysse, *La Création de l'Homme*, pp. 157, 184c. Gregory of Nyssa, *On the Making of Man*, *PG* 44, 184C.

74. Symeon the New Theologian, *The Ethical Discourses* vol. 1, 96–101, *SC* 122, p. 230.

75. Symeon the New Theologian, *The Ethical Discourses* vol. 10, 448–470, *SC* 129, p. 292.

76. Nicholas Kabasilas, *On the Life in Christ*, 4, *PG* 150, 601B.

77. Nicholas Kabasilas, *On the Life in Christ*, 4, *PG* 150, 641D.

78. Symeon the New Theologian, *The Ethical Discourses* vol. 10, 188–192, *SC* 129, p. 272.

79. Maximus the Confessor, *Mystagogy*, pp. 224–228.

80. John Damascene, *An Exact Exposition of the Orthodox Faith*, 4,13, *PG* 94, 1136–1153. Nicholas Kabasilas, *Commentary on the Divine Liturgy*, 22, *PG* 150, 146.

81. John Chrysostom, *Homily on Ephesians* 3,1, *PG* 62, 27–30. Vlasios Feidas, *Ecclesiology between Christology and Pneumatology in the Light of the Patristic Tradition*, pp. 52, 57.

82. John Chrysostom, *Homily on Ephesians* 4,16, *PG* 118, 1221. *Homily on Ephesians* 3,11, *PG* 62, 84. Epistle to the Romans 6,5.

83. John Chrysostom, *On the Ascension*, 16, *PG* 52, 789. *Homily on Hebrews*, 6,2, *PG* 63, 56. *Homily on Ephesians* 20,4, *PG* 62, 140.

84. Ioannis Karmiris, *Orthodox Ecclesiology, Dogmatics Part E'*, p. 164.

85. John 16,4. Acts 1,8. Peter 4,8. Acts 2,1. Gregory of Nazianzus, *Discourse*, 41,9, *PG* 36, 441.

86. Acts 2,38–41. First Epistle to the Corinthians 12,13. Luke 3,16. First Epistle to the Corinthians 10,16–17. Acts 8,14–17.

87. Cyril of Alexandria, *Commentary on John*, 11,11, *PG* 74, 560. Vlasios Feidas, *Ecclesiology between Christology and Pneumatology in the Light of the Patristic Tradition*, p. 52. First Epistle to the Corinthians 11,26. "You are Peter, and on this rock I will build my Church, and the gates of Hades shall not prevail against it," Matthew 16,18.

88. Cyril of Alexandria, *On First Corinthians*, 12,7, *PG* 74, 885.

89. John Chrysostom, *Homily on the Holy Pentecost*, 1,2, *PG* 50, 456.

90. The Apostolic Canons, 34, *Library of the Greek Church Fathers and Ecclesiastical Authors*, vol. 2, Apostoliki Diakonia (Athens: 1955), p. 175.

## *Chapter 12*

# The Orthodox Christian Ethics

Orthodox Christian ethics' main source is divine Revelation, as recorded in the Old and New Testament, and the Tradition of the Church, as interpreted by the Patristic Literature and the Church's experience.

Christian ethics involves awareness of God's will and, simultaneously, human compliance with it. Moreover, Christian ethics' linchpin is Revelation; it examines the rules related to free will, from the perspective of human likeness to God. The moral teaching has a theological and doctrinal character based on faith in the Triune God, following the model of Jesus Christ's life.[1]

The Revelation of the Triune God is the source of morality described in Scripture and in the Tradition of the Church, as memorandums of God's Revelation, offering elements for the development of Christian morality.[2]

The ethos of the Church is shaped by the texts of the Scripture, the Old and New Testament, the Sacred Tradition of the Church, theology, the Holy Canons, the Ecclesiastical Literature, the liturgical life, and texts, as well as by the personal experience of the Saints as Christ's emulators.

The principal norm of morality and provider of the moral law is God, being supreme wisdom and infinite benevolence,[3] the perfection and truth: "be perfect as your heavenly Father is perfect"[4] and "I am the truth."[5] In relation to humans, morality is determined by their compliance with the will of God and therefore by holiness, "Be holy, for He the Lord your God is holy."[6]

Ethics, based on doctrinal teachings, examines and demonstrates how humans will achieve, as much as possible, likeness to God. Humans will accomplish this through their relations with God and through their theoretical and practical stance and actions, to familiarize themselves with salvation as a gift from God.[7] Orthodox Christian ethics is directly connected to the doctrinal teachings of the Church and strives for the perfection of humans in the image of God.[8]

The theological starting point of the Orthodox Church's ethics is Trinitarianism and Christology. From Trinitarianism, it deduces material regarding motives for behaviors and actions, such as the loving community, the unity of will and action, and sociability. From Christology, ethics is defined by the teachings and life of Jesus Christ, and by how he dealt with and addressed the institutions of his time, for example, political power, the family, the Law.[9]

Furthermore, in Patristic Literature, ethics is described as the general Christian ethos with the new interpretation of classical education. Ethics has a Christocentric character, despite any references or influences from classical philosophers. Jesus Christ, in His human nature, is the ultimate standard for a moral teacher, defining human relations with God and the world.[10] Morality is based on faith in Christ and the cultivation of conscience considering His commandments.[11] Virtue has a Christocentric and ecclesiological character with an eschatological perspective.[12]

Christ is the perfect God and perfect human, the new Adam whom the Creator has blessed, so that the ethos of Christ is the ethos of the perfect human.[13] The Church pastors are partners in the freedom of the believers, in their pursuit of perfection and salvation.[14]

The doctrinal teachings of the Church—apart from the Trinitarian, Christological, and Holy-Spiritual expression—are consistent with the believers' life in Christ, on their journey towards perfection and theosis-deification, expressing the true ethos of a believer's life.

Christ is the yardstick of human morality, in which, through His sacrifice, He manifests humility, love, and forgiveness as an example for every believer to follow. By emulating the ethos of Christ, the believer partakes in the sacrifice of the cross and the expectation of the Resurrection within the body of the Church.

Ethics is the expression of God's will as the provident and creator of the world, and therefore ethics is determined by agreement and dependence on the will of God, on the new life in Christ. The uncreated God is the beginning of everything, who, being good, is the origin of morality, and, as possessing intellect, is the origin of reason.[15] Christian morality holds the perspective of the Resurrection through Jesus Christ by the grace of the Holy Spirit.

Morality is directly related to the virtues, since, apart from the effort and practice of the believer, it is at the same time the fruit of the Holy Spirit.[16] Virtues are fruits of the Holy Spirit[17] and characteristics of the human born again in Christ, presupposing the alignment of human will with that of God.

Love unites humans with each other and with God.[18] Virtues and their practice, in both their moral and spiritual dimensions, as fruits of the Holy Spirit, are the characteristics of humans' likeness to God.

Thus, Christian ethos, through an ecclesiological approach, is defined and referred to in relation to humans, the relationship of the believer with the community or society, human dignity and freedom, the concept of love, peace, and justice, and the understanding of equality.[19]

The communion of the created humans with the uncreated God takes place in the Church. God is the source of grace and virtue, and humans are becoming like Him, by participation of His virtue and benevolence.[20] The healing of humanity is achieved through the theoretical and practical application of morality in conjunction with the Sacraments of the Church.

Morality has an ecclesiological and sacramental dimension, as the believer shapes their ethos within the Church, the body of Christ, through participation in the Holy Sacraments. The intent is the renewal of humans from a pedagogical perspective through the Divine commandments, which determine the ethos of every believer and are consummated by their participation in the Holy Sacraments.

Morality is recapitulated in the life in Christ, which is experienced through the Sacrament of the Eucharist.[21] Through this sacramental union with Christ, morality is directed towards the spiritual perfection and theosis-deification of humans.[22] The alignment of human will with the volition of God brings the grace of the Holy Spirit into the life of the believer, culminating in the Eucharistic communion, thus achieving the connection and sacramental union of the individual with Christ and His body, the Church. During the Holy Liturgy, the unity of humans with the Triune God is affirmed, highlighting the divine ethos of the Father's love, the humility of the Son, and the communion of the Holy Spirit.

Christ's ethos is based on the freedom of conscience, without imposition: "For, brethren, ye have been called unto liberty; only use not liberty for an occasion of the flesh, but by love serve one another."[23]

A person's conscience is shaped by their relationship with God and the world. Conscience refers to the volition of God, existing within as an inherent moral law and acting as an inspirer and a teacher.[24] The divine law, innate in a healthy conscience, by adhering to the commandments, draws one closer to God as the source of goodness and holiness.

A healthy conscience is fostered by the promotion of correct values, accompanied by corresponding actions, from which emerges the concept of "having a good conscience."[25] But Christian moral conscience is determined solely by the Word of God, Jesus Christ.

The Apostle Paul speaks of "faith and a good conscience."[26] Conscience is directly related to human spiritual orientation and is shaped by the Holy Revelation and God's commands.[27]

Just as an anthropological approach considers a person as a single psychosomatic entity, so in theology, the moral and spiritual life are regarded as inseparable and unified.[28]

The Christian ethos is shaped by adherence to the divine commandments. The words and will of God are uncreated Divine actions, as opposed to the created nature of human words and actions.[29] Thus, the harmonization of human will and actions with the divine will makes one a participant in the uncreated divine actions, cultivating the divine ethos.

By keeping the commandments, the will of the believer under God is activated so that the rational will is attuned to the natural will, and a person can return to his natural state.[30]

Virtue refers to humanity's relationship with God since God is the source of all virtues. The more a human applies the will of God and cultivates a personal relationship and communion with Him, the more he is on the path toward His likeness.[31]

Since humans are created by God in His image and possess the potential for likeness, they inherently have natural virtues such as valor, wisdom, prudence, and benevolence.[32] For these virtues to be activated and bear fruit, they must be fortified by the Holy Spirit, enabling them to take on a spiritual character and enlighten the soul.[33] Thus, the attainment of virtue is the result of the grace of the Holy Spirit combined with human effort, with virtues being the fruits that assist and sanctify the believer.[34] These virtues are interdependent, with the cultivation of one strengthening the others, especially discernment and love. Virtue influences all aspects of human life—thoughts, words, and deeds—which are given meaning through life in Christ.

The principle of virtue is moderation, seen as the golden mean[35] between excess and deficiency, with spiritual implications relating for human relationships with God and others.

The Sermon on the Mount with the Beatitudes epitomizes Christian ethics, emphasizing virtues such as humility, God-inspired mourning, meekness, justice, mercy, purity of heart, peacemaking, as well as a stance against injustice.[36]

God's commandments are not mere moral precepts but manifestations of Divine actions, in which God is present. Hence, observing them is a testament to the encounter, personal relationship, and communion with Christ. Through the free observance of the commandments, human moral renewal is realized, culminating in participation in the Sacraments of the Church, particularly the Eucharist. Observing the commandments fosters the cultivation of virtues directed towards fulfilling God's will, allowing the believer to partake of God's holiness and grace, for God's commandment is "eternal life."[37]

Repentance in the life in Christ is the foundation and essence of Christian morality and life.[38] Through repentance, the believer's life is transformed, allowing the moral to take on a spiritual character, accompanied by appropriate actions.

According to John of Damascus, "prayer is lifting up your mind to God."[39] John Sinaites views prayer as human reference to, communion with, and union with God.[40] In prayer, a person is likened to a sunflower that follows the path of the sun. God, as light,[41] illuminates the individual who turns to Him, nourishes and enlivens him spiritually.

Prayer, as the mother of virtues, elevates humans to God, giving birth to virtues and rendering the individual godlike.[42] Through continuous prayer, the mind and heart are enlightened, virtues are strengthened, and humans become "temples of the living God."[43] Holiness is the objective and the pinnacle of human moral perfection, echoing the evangelical commandment "Be holy, for I am holy."[44]

A person's holiness arises from the possibility and the encouragement of imitating Christ, which derives from his incarnation.[45] Thus, the believer is called upon to imitate Christ "to the degree of the incarnation"[46] and by "being a person within their power of assimilation to God.[47] Ethics assumes a purely Christological dimension, given that the believer's love for Christ manifests itself when they imitate him within their power.[48]

Also, just as the saints imitated Christ and their morality took on a purely holy-spiritual character, so too the believers are asked to imitate the Saints, and according to Apostle Paul, "be imitators of me, as I also am of Christ,"[49] and according to the Gospel, "Be ye perfect, even as our Father who is in heaven is perfect."[50]

The ascent of humans on the scale of virtues bears fruits by human's peacemaking with themselves, their fellow individuals, and God. The bond of peace transforms the believer into the Son of God, according to the relevant Beatitude.[51] It is the first greeting of the risen Jesus to his disciples, proclaiming a new age for those who keep his commandments. Cultivating the virtues, with Jesus' death on the cross and his Resurrection as its main point of reference, unites people in a new humanity where the God of love and peace reigns.[52]

## THE IMPORTANCE OF LOVE

The three stages of believers' perfection correlate with the theological virtues of faith, hope, and love with the third stage being theosis-deification, ruled by the virtue of love.[53]

John the Evangelist underscore the significance of virtue and love in Jesus' public ministry.[54] Love, as a virtue, is theocentric, originating from God. The reciprocal love between God and humans also shapes the love among humans. God presents Himself to humans as a God of love. The intra-Trinitarian unity of love, as expressed in Christ's high priestly prayer,[55] reflects the love between the persons of the Holy Trinity and is experienced within the Church community, laying the foundation for salvation.

Love is the "new commandment," illustrating divine love and instructing Jesus' disciples and all believers to emulate this love. The disciples' love for Jesus is also manifested through their mutual love "with one another,"[56] forming a loving community united by the love for Jesus Christ. This regenerative love demonstrates that those who practice it know God and have been reborn by Him, while also setting the disciples apart from the world.[57]

Love, the highest virtue, relates to true observance of the commandments and aligns with Jesus' teachings; it encompasses healing the sick and welcoming sinners, even the rejected,[58] inside the divine fold. This agape, with its ecclesiological aspect, imitates the love in Christ, signifying the commencement of new life rooted in "I give you a new command to love one another, as I have loved you, so you must love one another,"[59] according to the dictum of divine love: "this is my commandment, that you love one another as I have loved you."[60]

The love of the Father, the grace of Jesus Christ, and the communion of the Holy Spirit become the lived legacy within the Church, fulfilling God's commandments. Demonstrated love verifies a person's love for God, consummated by following His commandments.[61] This lived connection within the Church is rooted in "love in one another," "all in one" in a communion of faith,[62] resulting in salvation through the love shared in Christ.

The Father's love is revealed through the presence of the Son and Word in the world,[63] who sacrificed himself for the salvation of man, embodying the divine love.[64] The sacrifice of the God-man Jesus Christ on the cross is the testimony of the love of God the Father[65] and of Christ's love for humans, defining the love of the disciples for their fellow human beings: "love one another as I have loved you, no one has a greater love than this."[66]

The Triune God is love, and thus love is the foundational element of all community. Since God is the source of love, humans partake and commune with the love from God.[67] Through love, the inner unity and the communion with the fellow human and God are maintained, since "God is love and whoever lives in love lives in God, and God in him."[68] Symeon the New Theologian describes love as "the communion and participation of divinity."[69]

In Orthodox Tradition, love is linked with dispassion, as through dispassion, sinful habits wane and divine passion flourishes, fostering other virtues.[70] When dispassion and love are experienced, humans transcend passion's tyranny, treating all individuals equally.[71]

Within the Church, love is not a mere concept but an ontological reality, reflecting the nature of the Triune God, for "God is love and whoever lives in love lives in God, and God in him."[72]

Theologically, love characterizes the intra-Trinitarian relationship as the uniting force among the three persons. On the human level, God's love manifests in His creative, providential, and salvific actions. God's love for humans is evident in the incarnation of the Word. Henceforth, salvation and new life are intrinsically linked to life in Christ, evidenced by keeping His commandments.[73]

Love is coupled with faith and hope, the triad of theological virtues.[74] Love extends from faith and hope towards recognizing God and accepting His Son as the mediator and savior. Through love, humans become disciples and friends of Christ,[75] directed both towards God and others.[76] As per John the Evangelist, the transition from death to life is through love for the brethren.[77]

God's initial love for humanity, demonstrated by sending His Son to save the world,[78] invites humans to participate in the redemptive act of divine love.[79]

Love for God integrates humans into the love shared between the Father and the Son in the Holy Spirit. Given this loving intra-Trinitarian relationship, those who love God share in holy love as disciples of Christ, united with Him through the Holy Spirit.[80]

Love through Jesus Christ dispels fear, as it is the perfect love.[81] Participation in divine love transforms humans by grace and is the essential prerequisite for loving others, embodying perfection, and unity. Love is fortified and sustained through prayer. The Church Fathers emphasize prayer's role in spiritual life, with references throughout the Old and New Testament.[82]

In prayer, believers express their entreaty, loving disposition,[83] and gratitude[84] to God, within the context of communication with a faithful and devotional aspect.

Love has a charismatic character and is realized in Christ within the Church in a unity of peace.[85] Eros is the "intense" love,[86] which, according to Dionysius the Areopagite, is a unifying and binding force that brings people into communion and restores communion with God.[87]

## NOTES

1. Ioannis Karmiris, *Christian Ethics*, pp. 6, 7.
2. George Mantzaridis, *Christian Ethics I, Introduction-General Principles, Modern Challenges* (Pournara, Thessaloniki: 2004), p. 79.
3. Karmiris, *Christian Ethics*, p. 35.
4. Mark 10,18.
5. John 14,6.
6. Leviticus 9,2. Matthew 6,10. John 4,34. Epistle to the Romans 12,2.
7. Karmiris, *Christian Ethics*, p. 22.

8. Matthew 5,48.
9. Apostolos Nikolaidis, *Social Ethics*, Grigoris (Athens: 2009), p. 78.
10. Panagiotis Dimitropoulos, *Orthodox Christian Ethics* (Athens: 1970), p. 5.
11. Luke 3,22. George Mantzaridis, *Christian Ethics I*, p. 67.
12. Mark the Hermit, *Of the Spiritual Law*, 196, *PG* 65, 929A. Gregory of Nyssa, *On Perfection*, *PG* 46, 285C.
13. Mantzaridis, *Christian Ethics I*, p. 74.
14. John Chrysostom, *On the Priesthood*, 2,3, *PG* 48, 634.
15. Clement of Alexandria, *Stromata*, 4,25, 162,5.
16. Galatians 5,22–3.
17. Galatians 5,22.
18. Second Epistle to the Corinthians 5,14.
19. Apostolos Nikolaidis, *Social Ethics*, p. 78.
20. Maximus the Confessor, *Various Chapters on Theology and Economy*, 1,50, *PG* 90, 1101AB.
21. George Mantzaridis, *Christian Ethics I*, p. 56.
22. Mantzaridis, *Christian Ethics I*, p. 69.
23. Galatians 5,13.
24. George Mantzaridis, *Christian Ethics II, Man and God, Man and Fellow Man, Existential and Bioethic Propositions and Perspectives*, pp. 42, 43. Epistle to the Romans 2,14–6.
25. 1 Peter 3,16.
26. First Epistle to Timothy 1,19.
27. Ioannis Karmiris, *Christian Ethics*, vol. A' (Athens: 1967), p. 97. Panagiotis Dimitropoulos, *Orthodox Christian Ethics*, pp. 58, 59.
28. George Mantzaridis, *Christian Ethics I*, pp. 46, 47.
29. Archim. Sophrony Sakharov, *St. Silouan the Athonite*, Monastery of Saint John the Baptist (Essex, England: 1999), p. 283.
30. Maximus the Confessor, *On Various Questions*, *PG* 91, 1048B.
31. Gregory of Nyssa, *Homily on Prayer*, 4, *PG* 44, 1165AB.
32. Gregory Palamas, *Homily 3*, *PG* 151, 37A.
33. Symeon the New Theologian, *Catechisms*, 33,10–16, J. Darrouzes, *SC* 113 (Paris: 1965), p. 248.
34. Gregory of Nyssa, *On the Divine Purpose*, *PG* 46, 289C. Galatians 5,22–3.
35. Maximus the Confessor, *Disputations with Pyrrhus*, *PG* 91, 309C.
36. Matthew 5,3–12. Luke 6,20–49.
37. John 12,50.
38. Gregory Palamas, *Homily* 59,1, ed. Oikonimou, p. 235.
39. John Damascene, *PG* 94, 1089C.
40. John Sinaites, *The Ladder*, 28, *PG* 88, 1129A.
41. Hebrews 12,29. Second Epistle to the Corinthians 4,6.
42. Mark the Hermit, *Concerning Those Who Think to Be Justified Through Works*, 33, *PG* 65, 936B.
43. First Epistle to the Corinthians 6,19.
44. 1 Peter 1,16.

45. Leontios Byzantios, *Against Nestorius and Eutyches*, 2, *PG* 86, 1349A.
46. Basil the Great, *Conditions by Amplitude*, 43, *PG* 31, 1028BC.
47. Clement of Alexandria, *Stromata*, 1,11,52.
48. Maximus the Confessor, *Chapters on Love*, 4,55, *PG* 90, 1060C.
49. First Epistle to the Corinthians 11,1.
50. Matthew 5,48.
51. Matthew 5,9.
52. Luke 2,14. First Epistle to the Thessalonians 5,23. Second Epistle to the Corinthians 13,11.
53. Maximus the Confessor, *Various Chapters*, 5,93, *PG* 90, 1388CD.
54. John 1,12.13–17.
55. F. Konstantinos Papathanasiou, *Salvation in Johannine Theology* (Pournara, Thessaloniki: 2010), p. 564.
56. Papathanasiou, *Salvation in Johannine Theology*, p. 585. Savvas Agouridis, *John the Evangelist, Introductory, Explanatory and Theological Studies on the Fourth Gospel* (Athens: University of Athens, 1975), p. 14.
57. Papathanasiou, *Salvation in Johannine Theology*, p. 585.
58. Agouridis, *John the Evangelist, Introductory, Explanatory and Theological Studies on the Fourth Gospel*, p. 14.
59. John 13,34.
60. John 15,12.
61. Papathanasiou, *Salvation in Johannine Theology*, p. 585. John 14,15: "If you love me, keep my commandments."
62. John 17,21.
63. Papathanasiou, *Salvation in Johannine Theology*, p. 585.
64. Papathanasiou, *Salvation in Johannine Theology*, p. 585. 1 John 3,16–17
65. 1 John 4,11.
66. John 13,34. 15,12.
67. Nikolaos Matsoukas, *Cosmos, Man, Society according to Maximus the Confessor*, Grigoris (Athens: 1980), pp. 255, 256.
68. 1 John 4,16.
69. Symeon the New Theologian, *Ethical Discourses*, 4,556, ed. J. Darrouzes, *SC* 129 (Paris: 1965), p. 248.
70. John Sinaites, *The Ladder*, 29, *PG* 88,1149A. Mantzaridis, *Christian Ethics II, Man and God, Man and Fellow Man, Existential and Bioethic Propositions and Perspectives*, p. 69.
71. Maximus the Confessor, *Chapters on Love*, 2,30, *PG* 90, 993A.
72. 1 John 4,16.
73. 1 John 5,3. 2 John 6.
74. First Epistle to the Corinthians 13,13.
75. Dimitropoulos, *Orthodox Christian Ethics*, p. 126.
76. Deuteronomy 6, 5.10, 30. Mark 12,30. Luke 10,27. Mark 12,31.
77. 1 John 3,14.
78. 1 John 4,20. John 3,17.
79. Dimitropoulos, *Orthodox Christian Ethics*, p. 126.

80. Dimitropoulos, *Orthodox Christian Ethics*, p. 127. John 14,23. 4,16.
81. 1 John 4,8. 16,18.
82. Luke 11,1. Matthew 4,1. Epistle to the Romans 8,26. Galatians 4,6.
83. John 16,23. Matthew 7,8.
84. Galatians 4,2.3.
85. Mantzaridis, *Christian Ethics II*, p. 333.
86. Gregory of Nyssa, *On the Song of Songs*, 13, *PG* 44, 1048C.
87. Dionysius the Areopagite, *On Divine Names*, 4,15, *PG* 3,713A. Mantzaridis, *Christian Ethics II*, p. 363. Vladimir Lossky, *The Mystical Theology of the Eastern Church* (Crestwood, NY: St Vladimir's Seminary Press, SPS, 1976), p. 113.

## *Chapter 13*

# The Soteriological Approach According to the Orthodox Theology

According to the Fathers, the salvation, redemption, and theosis-deification have a trinitarian character, since they are gifts from the three persons of the Holy Trinity. In particular, the participation of the Holy Spirit is directly related to the presence of God the Father and God the Word. Father reveals himself, acts, and saves through the Word in the Holy Spirit.

The salvation of humanity is the work of the Triune God, and when one person of the Holy Trinity acts, the whole Trinity works together. The salvific work of the incarnation of the Son and the Word is at the same time a labor of the love and favor of the Father and of the energy and grace of the Holy Spirit: "the Father, the Son and the Holy Spirit have distributed the economy which is for us."[1]

The salvific actions, the gifts, the donations, and every good thing derive from the Holy Trinity and as "a move and propagation of the benevolent made, coming from the Father through the Son to the Spirit."[2]

As the Divine Economy unfolds, a gradual revelation of the Son and Word of God in history takes place. First with the incorporeal revelation of the second person of the Holy Trinity in the Old Testament and then with his incarnation in the New Testament. The purpose of the work of the Divine Economy and the Church, as His body, is the renewal and salvation of man. Man's salvation is rooted in the divinity of Jesus Christ, who as the God-man reveals the one Triune God to the world, through His common providential and salvific actions.[3]

This theocentric aspect of man's salvation is accomplished through Christ in the Holy Spirit. The incarnation of the Word takes place in the Holy Spirit and the Holy Spirit continues its work within the Church, which is the mystical body of Christ. It is under the Divine Economy, that the Holy Spirit works within the Church for the attainment of man's salvation.[4]

According to Athanasius the Great, the salvation of humanity has a Trinitarian dimension. For the salvation of man, in the context of the Divine Economy the Father conceives and blesses the work of creation and redemption of man, the Son Word carries out the Father's will and the Holy Spirit concludes the work of theosis-deification and salvation by its sanctifying action.[5] The mystery of the incarnation of the God Word is regarded as the work of the Holy Trinity, in the context of the Divine Economy, which works towards the salvation of man.[6]

Because of Adam's disobedience, corruption was introduced into human nature and consequently into the entire human race, but with the incarnation of the Son of God, human nature was restored in its entirety. Jesus Christ renewed human nature and led humanity to salvation, to a regenerated unity of the whole man, but also to a true union of human with God in the person of Christ.[7]

The second Adam took over, united the human with the divine nature, and healed human nature because "the received is incurable."[8] In addition to healing, He regenerated, sanctified, and redeemed humanity from corruption and sin by becoming "the cause of eternal salvation."[9]

Humanity's renewed nature in Christ, as a "new creation," frees them from sin and recreates the new, through Christ, "regeneration towards sanctification in the Spirit,"[10] which leads to salvation and sanctification.

Humanity's regenerated nature, through Jesus Christ, overcomes decay and death, as the Son and Word, through the assumption of human nature, granted salvation, incorruption, and life.[11]

The Word of God, during the Incarnation, realized humanity's salvation and, by assuming a human body without sin, condemned corruption.[12] Christ embraced the indestructible passions of human nature and healed them, leading humanity to healing, apathy, and holiness for the salvation of the world.[13] The "New Man," Christ, through His "salvific passions," healed humanity, freeing them from corruption and sin.[14]

Christ, the new Adam, assumed human nature to eradicate decay and sin from it, to heal it and to grant it apathy and the victory over decay and death.[15]

Humanity's renewal in Christ does not merely restore human nature to the pre-Fall state of Adam but elevates it to the new state of the New Adam who overcomes sin and decay, granting immortality, holiness, and incorruption. Human nature is rescued and deified, taking on the characteristics of the uncreated nature by grace through Jesus Christ in the Holy Spirit.[16]

The incarnation of the Son and Word of God offers the possibility for theosis, immortality, and humanity's salvation, as expressed by Athanasius the Great: "For He became human so that we might become God . . . and He manifested Himself by means of a body so that we might perceive the unseen Father. And he endured human hybris so that we might inherit immortality."[17]

According to Cyril of Alexandria, through the incarnation of the Son of God, human nature was deified, and its salvation in Christ goes beyond restoration to the pre-Fall state. Salvation in Christ paved the way to perfection based on the restored image, so that all humanity may be led to the likeness of God and theosis.[18]

Human theosis and salvation are achieved through the incarnation of God the Word and are realized through the exchange of properties of the two natures in the person of Christ, during which the divine nature exchanges its properties with the human one: "give in return ourselves."[19]

The human and divine natures were united from the very beginning in one hypostasis of Christ, so that He is one: one person "with two natures," the divine and the human. The human nature, during the hypostatic union with the divine nature, was elevated, perfected, and deified by the assumption and union with the divine nature.[20]

After the Birth, the Passion, the Resurrection and the Ascension of Christ, humanity is recreated and saved by the Holy Spirit in a new life, as it is renewed and sanctified.[21]

Upon Christ's Resurrection, the human nature of Jesus is resurrected, leading to the redemption of human nature from death and decay. Humanity is led toward perfection. The Resurrection made the human nature of Christ the "dwelling place" of the Holy Spirit and, consequently, the "pathway" for the outpouring of divine grace to all creation. Jesus Christ became the "Last Adam," the "firstborn from the dead," the "beginning" of all humanity, the second beginning of "those on earth," the "Forerunner" of the path to heaven, the "road" to Resurrection and life.[22]

The salvation of human nature, besides the Passion on Golgotha and the Resurrection, is also accomplished by Christ's Ascension into heaven. Human nature achieves merit through both the ascent into heaven and the sitting at the "right hand of the Father," receiving the glory of sonship in the person of Jesus Christ.[23]

During the Ascension of the God-man Christ into heaven at the "right hand of the Father," the human nature of Jesus was elevated to a "transcendent glory"[24] and "placed us in the presence of the Father."[25] The ascension of human nature into heaven, carried in the person of the God-man Jesus, bestows on humanity sonship and salvation.

Following the Incarnation of the Word of God and the Ascension of Christ, human synergy is necessary for humanity's salvation and theosis. This synergy is directly connected with the spiritual birth in Christ within the Church, which occurs through the Sacraments, but also through the imitation of Christ, realized through the grace of the Holy Spirit.

According to the soteriology of Athanasius the Great, acknowledging the incarnation of the Son and Word of God in the person of Christ[26] constitutes the soteriological basis of humanity's synergy with Divine grace, leading to theosis.

In practice, humanity's synergy pertains to the upright faith, the living of a virtuous life as a genuine declaration of the upright faith that guides the believer to their personal salvation.[27] Humanity's synergy is assessed by the fact that human nature is subject to fickleness, with a tendency to be led either to alienation or nihilism, or to perfection or theosis. Any positive or negative alteration or change in humanity depends on their free will and self-authority.[28]

Humanity's freedom of choice is directly and indirectly related to fickleness, decay, or renewal.[29] The free human as a fickle and independent person, in combination with the gift of rationality based on their choices, possesses the ability to cultivate an interpersonal relationship with the Creator-God in a dynamic path that leads to perfection and salvation.

Depending on the free use of self-authority and following the dynamic and evolutionary perspective of the "in the image and likeness," humanity's choices can lead them to a path of perfection, with the participation of Divine actions, or vice versa, provided that the individual is free in his choices.[30]

Salvation is granted by Christ through faith, love, the practice of virtues because of the observance of the commandments, with the ecclesiological dimension of communion within the body of Christ, where the individual is perfected through the participation in the Holy Sacraments.

The incarnation of the Word of God has brought about the salvation of humanity and bestows the Holy Spirit through the Sacraments that take place in the Church as His body.[31]

Salvation is of a dynamic character, because the renewal and the communion of the individual, as a person, with his personal God face to face[32] is a relationship of free choice with the corresponding use of their self-authority. When the will and self-authority freely follow the Word of God, without coercion but of their own volition, to go along with the Divine commandments and will, then they partake of divinity, holiness, and are saved by God's salvific providence.

The salvation of humanity is accomplished within the sacramental body of Christ, which is the Church.[33] Within the Church, through the Holy Baptism, we get the joint sanctifying action of the Holy Trinity and by the grace of the Holy Spirit a person is directed towards perfection.[34]

According to Symeon the New Theologian, apart from the mental and by grace union of humanity, the tangible union of the believer with Christ, which takes place through the Sacraments, is also necessary.[35] The Sacraments are dual in character, tangible-visible and spiritual-mystical, just as humanity is dual, consisting of soul and body, since through the Holy Spirit the salvation of the whole human is accomplished within the body of the Church.[36]

Within the Church, the salvation of humanity and the world is worked out, through the union of God with humanity, which is proclaimed by the Son through the Holy Spirit.[37] The salvation of a person in Christ is realized through the action of the Holy Spirit. The communion of grace strengthens the bonds of communion of humanity's loving communion with God.[38]

Salvation calls for the synergy of a person's free will and the action of divine Grace.[39] The path that leads the believer to salvation, which is Jesus Christ, is the study of the Scriptures, which strengthens the faith, reveals the divine will, and encourages the believer to implement it, and thus, by keeping the divine commandments, the participation in the Kingdom of God is completed.[40]

The path to salvation is accomplished by the liberation from passions, God's love, and the absolute apathy on humanity's part, but it is completed by the assistance of the Triune God.[41] Salvation is predicated upon the observance of God's commandments in practice, since through this, the will of humanity becomes the will of God.[42]

Salvation paves the way for the healing and restoration of human nature. During this restoration, the energy and will of human nature conform to the uncreated divine energy and will. This alignment of the human energy and will with the divine bestows on humanity salvation and the sonship and theosis by grace,[43] since God grants life, love, incorruptibility, and immortality.[44] By the incarnation of the Son and God the Word, the old Adam is renewed, and the new Adam is given, according to the plan of the Divine Economy, the possibility of humanity's return "in accordance with nature" and salvation is gifted to humanity, making them immortal by grace. The Holy Spirit, the Paraclete, with its regenerating and sanctifying energy and power, is the active agent of the passing on of the divine gifts and energies.

The creation of humanity in the image and likeness of God is directly related to the mystery of the incarnation of the Word of God. Human nature as the image and likeness of God was the prerequisite for the incarnation of the Word of God and the union of God and humanity in one person, Jesus Christ, according to the Divine Economy. The reciprocation of the characteristics of the divine nature to the human nature of Christ healed humans and made them partakers of immortality, holiness, and divinity.

Moreover, following the incarnation of the Son and Word of God and by the regeneration and renewal of humanity with the New Adam, the new human of the Resurrection and the Ascension, the purified in the image and the possibility of likeness in Christ are the precondition for the possibility of humanity's theosis and salvation, as long as the union of the image with the Archetype and the likeness of humanity with the Holy God take place within the Church. Therefore, the image and likeness of humanity was the prerequisite for the incarnation of the Word of God and afterward the necessary element for the theosis and salvation of humanity within the body of the Church since the Church is considered the image and form of the Triune God.

Athanasius the Great claims that, since the creation is not composed of God's own essence, humanity can be saved by the participation of His divine actions, simply because God does not withdraw his salvific providence and life-giving actions.[45] Thus, humanity partakes of the divine actions, not the divine essence, and becomes a child by grace. Consequently, salvation according to the plan of the Divine Economy and the revelation of the Triune God is of a purely theocentric character.

With the Resurrection of Christ, the human race inaugurates the one life, the life of incorruptibility and holiness, with a purely eschatological perspective.[46]

The fall of humanity dragged down creation as well, but the salvific providence of the Triune God granted salvation within the charismatic body of the Church of Christ. Within this new body of Christ's Church, through the favor of the Father by the grace of the Holy Spirit, the salvation, and the possibility of the union of humanity and creation with God by grace are communed. This salvation is eschatological in character because creation and humanity are called to enter, through the Church of Christ, into the eternal heavenly Kingdom of God.[47] The salvific action of the Triune God has an eschatological perspective with cosmological, ecclesiological, and anthropological aspects.

## NOTES

1. John Chrysostom, *Homily on Pentecost* 1, 2, *PG* 50, 456. Ioannis Karmiris, *Synopsis of the Doctrine of the Orthodox Catholic Church* (Athens: 1960), p. 44.
2. Gregory of Nyssa, *Homily on the Romans*, 13,8, *PG* 60, 519.
3. Colossians 3,4. Gregory Palamas, *Homily XXI*, *PG* 151, 277AB. 1 John 5,1.
4. Athanasius the Great, *On the Incarnation Discourse against the Arians*, 12, *PG* 26, 1004B.
5. Athanasius the Great, *To Serapion* 1, 28, *PG* 26, 596ABC. Protopresbyter Christos Filiotis-Vlahavas, *The Teaching of Athanasius the Great on the Holy Spirit* (Thessaloniki: 2004), p. 167.
6. Panagiotis Dimitropoulos, *The Anthropology of Athanasius the Great* (Athens: 1954), p. 58.
7. Konstantinos Skouteris, *Man According to the Word of the Divine Economy*.
8. Gregory of Nazianzus, *Epistle to Cledonion* 101, *PG* 37, 181.
9. Hebrews 5,9.
10. Cyril of Alexandria, *Commentary on Isaiah*, 5, 2, *PG* 70, 1217A.

11. Cyril of Alexandria, *Commentary on Isaiah*, 5, 1, *PG* 70, 1188A. *Commentary on Luke*, *PG* 72, 908D–909A.

12. Cyril of Alexandria, *Commentary on Luke*, *PG* 72, 489D–492A.

13. Cyril of Alexandria, *On the Incarnation* 3, 26, *PG* 75, 1465B. *On Matthew PG* 72, 456C. *Commentary on John*, 7–8, *PG* 74, 92D.

14. Cyril of Alexandria, *On the Incarnation* 3, 28, *PG* 75, 1468B–C.

15. Cyril of Alexandria, *On the Incarnation* 3, 18, *PG* 75, 1452B–C.

16. Cyril of Alexandria, *Commentary on Isaiah*, 4, 5, *PG* 70, 1124A. *Commentary on John*, 10, 2, *PG* 74, 433B.

17. Athanasios the Great, *Discourse on the Incarnation of the Word and through the Body to Ourselves*, 54, *PG* 25, 192. Athanasios Delikostopoulos, *Greek Thought and Christian Conscience*, Apostoliki Diakonia (1993), pp. 271–281 [http: //www.apostolikidiakonia.gr/gr_main/catehism/theologia_zoi/themata.asp?contents=ecclesia_history/contents_texts.asp&main=texts&file=4.htm (15/02/20221)].

18. Cyril of Alexandria, *On the Incarnation* 3, 8, *PG* 75, 1425C–D.

19. Cyril of Alexandria, *Commentary on John*, 20,17, *PG* 74, 700. Athanasius the Great, *Against the Arians*, Discourse 2,70, *PG* 26, 296.

20. Ioannis Karmiris, *Synopsis of the Doctrine of the Orthodox Catholic Church*, p. 47. John Damascene, *An Exact Exposition of the Orthodox Faith*, III, 17, *PG* 94, 1068–1069.

21. Cyril of Alexandria, *Commentary on Luke*, *PG* 72, 501B. *Festive Homilies, Homily* 10, 1, *PG* 77, 609C. *Address to the Queens* 2, 46, *PG* 76, 1400B–C.

22. Cyril of Alexandria, *Commentary on Isaiah*, 3, 1, *PG* 70, 588B. *Commentary on Luke*, *PG* 72, 485A, 500B, 524B, *Commentary on John*, 2,1, *PG* 73, 228D. *On the Holy Trinity* 1,3, *PG* 75, 853B–C' 6, *PG* 75, 1008B–C, 1037D, 1044B. Chrysostomos Stamoulis, *Cyril of Alexandria, on the Incarnation of the Only-Begotten*, To Palimpsiston (Thessaloniki: 1998), pp. 126, 172.

23. Cyril of Alexandria, *Festive Homilies, Homily* 5, 7, *PG* 77, 497A.

24. Cyril of Alexandria, *Commentary on John*, 10, 2, *PG* 74, 432D.

25. Cyril of Alexandria, *Commentary on John*, 10, 2, *PG* 74, 432D.

26. Athanasius the Great, *Against the Arians* 1, 39, *PG* 26, 92C. Elias Moutsoulas, *The Arian Dispute and Athanasius the Great*, Part B', *Athanasius the Great*, vol. A' (Athens: 1978), pp. 189, 206.

27. Andreas Theodorou, *The Teaching of the Greek Fathers of the Church Up to John of Damascene on the Theosis-Deification of Man* (Athens: 1956), p. 237.

28. Nikolaos Matsoukas, *Dogmatic and Symbolic Theology B'*, p. 525.

29. Nikolaos Matsoukas, *Dogmatic and Symbolic Theology D', Satan* (Pournara, Thessaloniki: 1999), p. 77.

30. Christopher Eppling, *A Study of the Patristic Doctrine of Free Will* (Wake Forest, NC: 2009), p. 93. Nikolaos Matsoukas, *Theology, Ktisiology, Ecclesiology According to Athanasius the Great. Points of Patristic and Ecumenical Theology* (Pournara, Thessalonik: 2001), p. 72.

31. Symeon the New Theologian, *Hymn 55*, 142–147, *SC* 196, p. 264.

32. Exodus 33,11.

33. Symeon the New Theologian, *Hymn* 42, 48–55, *SC* 196, 40.
34. Athanasius the Great, *Against the Arians* 2, 41, *PG* 26, 236A.
35. Symeon the New Theologian, *Hymn* 30, 463–466, *SC* 174, p. 372.
36. Symeon the New Theologian, *Catechesis* 14, 122–123, *SC* 104, p. 212. Nikolaos Matsoukas, *Dogmatic and Symbolic Theology B'* (Pournara, Thessaloniki: 2001), p. 465.
37. Ioannis Karmiris, *Dogmatics Part E'*, *Orthodox Ecclesiology*, p. 79.
38. Cyril of Alexandria, *Commentary on Isaiah*, 4, 1, *PG*, 70, 888D–889A. *Commentary on John*, 9, 1, *PG* 74, 280A.
39. Ioannis Karmiris, *Synopsis of the Doctrine of the Orthodox Catholic Church* (Athens: 1960), p. 71. Clement of Alexandria, *Prophetic Utterances*, 22, *PG* 9, 708.
40. Symeon the New Theologian, *The Ethical Discourses* vol. 3, *SC* 122, 384–410, p. 418.
41. Symeon the New Theologian, *The Ethical Discourses* vol. 3, 394–396, *SC* 122. *Catechesis* 20, 14–20, *SC* 104, pp. 330–332.
42. Symeon the New Theologian, *The Ethical Discourses* vol. 3, 607–615, *SC* 122, pp. 432–434.
43. Nikolaos Matsoukas, *Dogmatic and Symbolic Theology B'*, p. 331.
44. Matsoukas, *Dogmatic and Symbolic Theology B'*, p. 532.
45. Athanasius the Great, *Against the Greeks*, 41, *PG* 25, 84A.
46. Cyril of Alexandria, *On the Veneration and Worship in Spirit and Truth* 17, *PG* 68, 1116D–1117A.
47. Vladimir Lossky, *The Vision of God*, p. 127.

## *Chapter 14*

# The Orthodox "Revelation" Theology

In accordance with tradition, the present life is organically linked to the future life, as the present leads to the future, and the future is founded on the present. Thus, the degree of communion with God in the present life provides the prospect and condition for its preservation in the future life, beyond the limits of time and space, extending to eternity.[1]

God is infinite and timeless, as He rules over space and time, being their creator.[2] The uncreated God connects with creation and acts within the created world, which testifies to the uncreated Creator. The creation of the tangible world, with humanity as its pinnacle, manifests the uncreated Creator. In addressing the manifestation of the Word of God from its eternal existence and its revelation to humanity, John the Evangelist associates the revelation of the Word with the renewal of humanity in Christ.

God, Jesus Christ, is "the Alpha and the Omega, the first and the last, the beginning and the end."[3] The perfection of everything is God, Jesus Christ, because "of him and through him and to him all things"[4] are the Kingdom and glory of God. During the incarnation, the unchanging Word of God, by the will of the Father, assumed created and mutable human nature to renew it, to animate it, and to make mortal humans "immortal, and to bring them into the eternal kingdom of heaven."[5]

Christ, through His advent in the world, renewed humanity and creation, placing His redemptive work within an eschatological framework of glory and anticipation: "We await new heavens and a new earth, as promised, where righteousness dwells.[6]

After death, which for humans results from the separation of soul and body, resurrection will follow as a reconnection of the soul with an incorruptible body. The Resurrection of the Lord signifies the beginning of the resurrection of mortals, wherein the body will be raised and made imperishable, and the soul reunited, fulfilling the prophecy, "Destroy this temple, and in three days I will raise it up."[7] At the Second Coming, humans will rise, their souls rejoining bodies now free from decay.[8]

As the pre-incarnate Word foreshadowed the incarnate Word, so too the incarnate Word foreshadows and heralds His future Second Coming.[9] Christology, delineating the works of the pre-incarnate and incarnate Word, underpins Ecclesiology, accentuating the historical and eschatological essence of the Church.[10]

In Jesus Christ, as a historical figure, the eternal intersects with the present and the end times, as He "is" pre-existent, "was" indicating His presence, and "is to come," the One who will return to exist in eternity.[11] The new Adam, Christ, signifies the beginning of the new Apocalyptic era within the Divine Economy, fulfilling the Prophets and the Law experienced in the Church, particularly through the Eucharist, which proclaims the end times and God's Kingdom present within the Church.

Christ is the God, "the one who is and who was and who is to come,"[12] with His glory manifest in the Church and His ultimate Kingdom. Every believer in the Church of Christ shares in the eschaton while living in the now. By uniting the divine with the human nature in Christ, the incarnation of the Son and Word of God bridges the now with the eschatological. Therefore, the Church's eschatological character as the body of Christ is defined through the expectation of Christ's Second Coming.[13]

As Christ's body, the Church exists eternally, "endless and infinite and the only one ever living" and "extends throughout eternity."[14] Also, "the Church is stronger than heaven, heaven, and earth will pass away,"[15] but the Church is everlasting.

The Church is Christ's body, and He is "the one who is and who was and who is to come."[16] His Resurrection implicates all humanity, as His suffering had the potential for universal salvation.[17] Christ is the "first-born of the dead."[18] He is the life and the Resurrection. God is the "place of the redeemed,"[19] where each person communes and partakes in eternal life according to their receptivity.[20] Christ, through his Resurrection, has bestowed eternal salvation and life.

In contemporary times, understanding Christ's witness through the Church and the anticipation of the eschaton is abandoned for political or scientific narratives.[21] The first part of this work described such views, often echoing the overconfidence of Babel and a neo-idolatrous glorification of technology. In contrast, the Church's life, integrating ethics with spirituality and the renewing grace of the Holy Spirit, heralds and prefigures the advent of God's Kingdom, already present among the Saints.

As Nicholas Kabasilas remarked, life in Christ begins now to carry on into the future.[22] This life unfolds within the Church, the mystical institution founded by God, where the conjunction of the present with the future and temporal existence with eternity is actualized.

The foundational tenets of the Church's eschatological perspective are human salvation, engagement in the Holy Sacraments with a focus on the Eucharist, and the deification and divine adoption of believers by grace. The Church's essence is intrinsically eschatological, as the faithful, through their active faith in divine commands, seize the present moment—viewed ecclesiastically as "kairos" (an opportunity for salvation)—to attain their ultimate salvific goal, which looks toward the future and the anticipated Kingdom of God.

The transforming power of the Church is beyond time and pertains to the past, the present, but above all to the prospect of the future and the renewal of humanity and all of creation.

God's revelation is summed up in the victory of Christ over death. The knowledge of the truth is beyond any intellectual action; it is the integration into the Church as the body of Christ and the conformity to His will by which the ontological renewal through Jesus Christ occurs.[23] The expectation of the Kingdom of God is experienced in the Church as already mystically present and at the same time as the future. The experience of the mystical presence of the Kingdom of God throughout history is experienced according to human receptivity. It is linked to anticipating the Second Coming of Christ as the end of history and the world.[24]

Baptism allows individuals to share Christ's death and resurrection to live in the reality of his risen life.[25] Christ's crucifixion serves as a beacon for human existence, enabling believers to participate in the life-giving power of His resurrection.[26]

Through the sacrament of the Eucharist, the believer is nourished and reborn spiritually as an organic member of the body of Christ, the Church.[27] God's self-disclosure, initiated through the Incarnation, continues through participation in the Holy Liturgy and Eucharist, where one encounters God's Kingdom.

The eschatological significance of the Church's sacraments is affirmed by Christ's resurrection and ascension. Christ's path paves the way for believers' resurrection and ascent to heaven. Partaking in the Eucharistic body and blood of the risen Lord within the Church attests to the sacramental completeness of union with the eternal and uncreated God, reflecting an eschatological reality and a hopeful anticipation.[28] The Eucharist, as the partaking of Christ's risen body and blood, sustains immortality, incorruptibility, and resurrection, epitomizing eschatological communion with immortality and resurrection.

Nicholas Kabasilas posits that through sacramental participation and practice, believers attain spiritual wellness in this life and purify their spiritual senses to anticipate the life to come.[29]

Basil the Great concurs, suggesting that within the ark of the Church, the faithful transfers his mind "from the present to the future," partaking in God's Kingdom.[30]

The Church's eschatological orientation is characterized by the believers' participation in divine glory and their experience of the Kingdom of God through spiritual senses. Through prayer and purity, they live a charismatic and eschatological life within the Church, notably via the Eucharist, communing with Christ's body and blood.[31] This imbues earthly life with a profound significance, for it is interwoven with the eschaton's renewing power, granting temporal actions eternal weight.[32] The Church thus stands as the threshold of the present and the future age,[33] offering believers the possibility of divine sonship from this world with an eternal horizon, all stemming from God's love and benevolence.

The Church is simultaneously an image of God and a microcosm of the universe, where the earthly and heavenly, material and spiritual, human and divine converge. Within the Church, creation is transfigured by uncreated divine glory, promising an eschatological future fortified by the incarnation of the Word and the grace conveyed through the Sacraments.[34]

According to Nicholas Kabasilas, the believers, as members of the Church, are also members of the body of Christ. The head of the body is concealed in the present age but will be revealed in the age to come. Then, when the glory of the head shines, the glory of the body's members will also shine.[35]

Incarnate in Christ, the immortal Word begets spiritually within the Church as the new Adam, who is to be immortal and incorruptible.[36] For whoever imitates the passion of Christ receives salvation "in truth."[37] The Church is the ark of truth, grace, and salvation, within which the eternal God saves humans "until the end of the age"[38] and "the gates of hell shall not prevail."[39]

Church life is a journey towards spiritual perfection and a foretaste of paradise stemming from the participation of God.[40] The spread of the Gospel prepares humanity for Christ's glorious return. It is then, at the Second Coming, that God and Saviour Jesus Christ will appear in the sky in all His divine glory.[41]

Thus, the Church's historical and eschatological journey is rooted in Christology. In Christ, the future penetrated history, merging the temporal with the eternal, physical with spiritual, heralding the eschaton.[42] For example, the event of the Lord's transfiguration has a purely eschatological character as a spiritual experience and foreshadowing of eschatological glory.[43]

The eschatological prospect is also experienced during the worshipping life in the doxology of the Trisagion Hymn, which has reference to the divine doxology of the Holy Angelic Powers of heaven, since in the future life, the participation of the incorruptible body of believers in it will also be realized.[44] According to the Church's teachings, this eschatological prospect has a Glorified character, with the main feature being the hope of Christ's coming in divine glory.

The Gospel's ethics, as the revealed Kingdom's ethics, lead through life in Christ to new creation communion. This new ethics reflects the new creation and resurrection, manifested in the world and experienced within the Church.

Divine Revelation guides humans to moral perfection and divine likeness.[45] The Lord's commandments and words prepare believers for His eternal Kingdom, with a direct link between moral perfection and blessedness. The Saints already partake of ultimate blessedness and foresee the future age through spiritual God's vision.

The "day of the Lord" prompts preparation, recognizing the present age's transience, liberating from worldly bondage, and directing towards God's eternal Kingdom.[46] The central point of repentance in its soteriological perspective has an eschatological character: "Repent, for the Kingdom of Heaven is near."[47] God raises those who repent above the tangible paradise to heaven, and grants them sonship by grace.[48]

Christ, bringing the Kingdom to people, taught that His words bear dynamic force; accepting His words is tantamount to embracing His Kingdom.[49] The Sermon on the Mount and the Beatitudes, with their eschatological nature, exalt the eternal's value and the new creation's promise.

The last day of the Lord's Revelation will be fulfilled, according to his promise, when his forbearance for the salvation of most is concluded.[50] The "day of the Lord" is the day of his final and complete revelation; it is the Lord himself; it is the day when the visible will recede, when he will be the "only one day and together God," when the time will extend alongside him.[51] The elusive and eternal God will become the endless and inexhaustible joy of the Saints because, through his incarnation, he became the standard of judgment of the righteous and the wicked for salvation.[52]

The Kingdom of God was revealed with Christ's entry into history and the world, yet it is also anticipated at the culmination of both history and the world, heralding the commencement of the age to come. Thus, the eschaton is presently experienced, imbuing the present with an eschatological essence.[53]

Revelation is bifurcated into natural and supernatural facets. The former is evidenced by the impact of divine energies and capabilities, whereas the latter pertains to God's direct intervention in history and the world. In both revelation types, God is the prime mover. Consequently, the world serves as the conduit for natural revelation, and the unfolding Divine Economy, as depicted in the Old Testament, history, and the Church, represents supernatural revelation.[54]

Christian eschatological hope extends beyond the resurrection to include the renewal of the creation.[55] Since nature "groans and travails together"[56] due to the fall of humanity and has been affected by the aftermath of sin, the regeneration of humans has beneficial consequences for the whole creation. The unity of humanity and the cosmos in Christ forms a core tenet of Orthodox cosmology and anthropology, enabling eschatological transformation through human restoration in Christ.[57]

The "potential" unveiling of God's Kingdom within historical reality was evidenced at the Transfiguration, the Resurrection, and Pentecost. According to Gregory Palamas, this unveiling recurs with every manifestation of divine glory to the Saints.[58]

God reveals Himself to humanity in myriad manners: through meditation, logic, observing creation, divine emotions, and through the Saints' experiences.[59] The Saints savored the revelations and the beholding of the Mysteries of God, after first overcoming their passions, toiling under God's commandments, and received the spiritual gifts as a reward.[60]

Following the commandments yields enlightenment in the present and assures a place in the "Kingdom of Heaven," simultaneously brightening the intellect to receive divine knowledge and revelation.[61] Paradise, both tangible and spiritual, is an extension of the divine life's Kingdom, where the vision of God is both eternal and ceaseless, thus dynamic and ever-progressing.[62]

Salvation is dynamic, initiated in this life through participation in God's uncreated energies and glories, prefiguring the eschaton. This participation renders humans a partaker of the future life in paradise, thus bestowing an eschatological dimension to human salvation and sanctification.[63]

Moreover, eschatology is marked by an apocalyptic nature with anthropological dimensions that transcend time, aligning with the scriptural proclamation, "The Kingdom of Heaven is within you."[64] The human heart becomes the vessel for the potential revelation of God's Kingdom, unfettered by temporal constraints.

## SIGHT AS A PERSONAL REVELATION

The relationship between the soul and the body, as a defining characteristic of humanity, and the act of creating humans as psychosomatic beings carry eschatological significance and references, according to the prophet Ezekiel.[65] In the resurrection, this psychosomatic unity will be realized.[66]

Thus, the purification of the mind and body and their sanctification are intimately linked to salvation, which entails combating passions and establishing the purified mind as the soul's sovereign. When cleansed of passions, the mind and body engage in virtuous practices.[67]

According to the patristic tradition, the praxis of practice, with the observance of the commandments and the cultivation of the virtues, is a prerequisite of "sight-divine vision," during which the mind experiences the highest spiritual state, which is illuminated by the uncreated light and ascends to the sight of God.[68]

An individual's eschatological outcome is shaped by their openness to God and their experience of His presence in this life, foreshadowing the richness of spiritual delights in heaven.[69] By grace and through observance of the commandments and participation in the Sacraments, believers transcend sin and death and step into eternity with God.[70]

With empirical knowledge of God, through vision, the human mind and heart partake of divine perfections, compelling Christians to emulate these perfections to become more godlike.[71] Theosis, therefore, has an eschatological dimension in which the unity of nature and communion with God is restored through the bond of love.[72]

The ascetic life and the practice of virtues create the conditions for divine vision, where the mind is bathed in uncreated light and elevated to a profound spiritual state—the vision of God as a foretaste of the life to come.[73]

Through the purification of mind and heart, one is warranted to direct the mind towards God with the heart's intrinsic illumination and behold God.[74]

Saints mentally perceive the Holy Trinity, for through intrinsic illumination, they become worthy of active divine engagement, enraptured by divine grace in a state of revelation and silent mysteries of the future.[75]

According to Nicodemos the Hagiorite, following the "sight of acceptance," God carries out the revelations and the view of the future externally and passively to those with a pure heart through the enlightenment of the heart and divine radiance.[76]

The purified and deified mind, graced by the Holy Spirit's energy, is granted visions of the age to come as it transcends all physical and intellectual creation.[77] The actions of the Holy Trinity are carried out by divine grace in the hearts of the purified ones, secretly and confidentially,[78] so that they may behold the perfections of God and the secret visions of the age to come.[79]

In such visions, the mind experiences joy, delight, and a foretaste of divine bliss and perfection, perceiving the future state as an anticipatory taste of promised blessings and paradise.[80]

Through this union with God-achieved via vision and divine sight-man, the image of God is conformed to and communes with the light and all divine virtues.[81] Gifts of discernment, insight, foresight, and the precognition of the future are imparted alongside the heart's intrinsic illumination, ecstatic divine love, and a partially concealed vision of divine mysteries of the coming days and the future Kingdom.[82]

The eschatological "eighth day" transcends time, eternal and without end, to be revealed as an everlasting and imperishable day[83] at the culmination of days.

This "eighth day," the Lord's undying day, correlates with the Holy Spirit's indwelling and the perfection of humanity through enlightenment and ascension to the divine, by grace, sharing in eternal joy, glory, and the Kingdom of God.[84] In this state of vision, God reveals Himself personally to humanity, bestowing by grace a share in divine glory, while the ineffable mysteries of the eschata, the forthcoming Kingdom of God, and the eternal day of the Lord are made known.

## NOTES

1. Gregory Palamas, *Homily 45,7*, ed. Oikonomou (Athens: 1861), p. 48.
2. Athanasius the Great, *Discourse on the Incarnation of the Word*, 17, *PG* 25, 125AB.
3. Apocalypse 22,13.
4. Epistle to the Romans 11,36.
5. Athanasius the Great, *Against the Arians* II, 65, *PG* 26, 305. Ioannis Karmiris, *Synopsis of the Doctrine of the Orthodox Catholic Church*, p. 43.
6. 2 Peter 3,13.
7. John 2,19. John Damascene, *An Exact Exposition of the Orthodox Faith*, Introduction, Comments, translation by Nikolaos Matsoukas, pp. 448, 449.
8. Damascene, *An Exact Exposition of the Orthodox Faith*, Introduction, Comments, translation by Matsoukas, pp. 452, 453.
9. Nikolaos Matsoukas, *Cosmos, Man, Society according to Maximus the Confessor*, pp. 248, 249.
10. Matsoukas, *Cosmos, Man, Society according to Maximus the Confessor*, pp. 247, 249. Ioannis Kalogeros, *On the Character of the Orthodox Catholic Church according to the Fundamental Soteriological Principles of the New Testament*, offprint of the honorary volume on the Jubilee of His Eminence Metropolitan of Philippi, Neapolis and Thassos Chrysostom (Athens: 1961), pp. 4–7.
11. Hebrews 13,8. "Behold, I am with you all the days until the end of the age," Matthew 28,20.
12. Apocalypse 1,4.
13. Maximus the Confessor, *Mystagogy*, p. 194.
14. John Chrysostom, *Commentary on the Psalms* 144, 4, *PG* 55, 469.
15. John Chrysostom, *Discourse on Blessed Babylas* 2, *PG* 50, 536.
16. Apocalypse 1,4.
17. Gregory Palamas, *Homily* 21, *PG* 151, 277AB.
18. Epistle to the Romans 8,29.
19. Maximus the Confessor, *Various Chapters on Theology and Economy*, 1,68, *PG* 90, 1108C. Maximus the Confessor, *Various Chapters*, 4,51–2, *PG* 90, 1328BC.
20. Maximus the Confessor, *Various Chapters*, 4,53, *PG* 90, 1328D.
21. George Mantzaridis, *Christian Ethics I*, p. 54.
22. Nicholas Kabasilas, *On the Life in Christ*, 1, *PG* 150, 439B.

23. Mantzaridis, *Christian Ethics I*, pp. 46, 47.
24. George Mantzaridis, *Christian Ethics II*, p. 117.
25. Epistle to the Romans 6,3–5.
26. Basil of Caesarea, *On the Holy Spirit*, 15, *PG* 32, 132A.
27. John 15,1–8. 6,54–56.
28. Stefan-Lucian Toma, *The Patristic Tradition in the Work of Father Dimitru Stăniloae and the Modern World* (Pournara, Thessaloniki: 2007), p. 236.
29. Kabasilas, *On the Life in Christ*, *PG* 150, 493–496. Nikolaos Matsoukas, *Dogmatic and Symbolic Theology C'*, pp. 280, 281.
30. Basil of Caesarea, *On the Holy Spirit*, 66, *PG* 32, 192C.
31. Mantzaridis, *Christian Ethics II*, p. 119.
32. Mantzaridis, *Christian Ethics II*, p. 120.
33. Mantzaridis, *Christian Ethics II*, p. 112. Luke 17,21.
34. Maximus the Confessor, pp. 230–240.
35. Kabasilas, *On the Life in Christ*, 2, *PG* 150, 548BC.
36. Symeon the New Theologian, *Ethical Discourses*, 1,10,160–165, *SC* 122, p. 174.
37. Cyril of Jerusalem, *Mystagogical Catecheses*, 2,5, *PG* 33, 1081A.
38. Matthew 28,20. John 14, 16, 17.
39. Matthew 16,18.
40. John Damascene, *Against Manichaeus*, *PG* 94, 154D–1548A.
41. Titus 2,13. Matthew 24,14. Apocalypse 1,4.
42. F. Nikolaos Loudovikos, *A Eucharistic Ontology* (Athens: Domos publications, 1992), p. 16.
43. Dimitrios Tselengidis, *The Soteriology of Luther* (Pournara, Thessaloniki: 1991), p. 242. Loudovikos, *A Eucharistic Ontology*, pp. 152, 153.
44. Maximus the Confessor, *Mystagogy*, p. 238.
45. Ioannis Karmiris, *Christian Ethics*, p. 177. Matthew 5,48.
46. Gregory Palamas, *Homily* 19, *PG* 151, 261D. Matthew 5,3, "Blessed are the poor in spirit, for theirs is the kingdom of heaven." Matthew 5,10, "Blessed are they which are persecuted for righteousness' sake: for theirs is the kingdom of heaven."
47. Matthew 3,2.
48. Hannah Hunt, "Penthos and Repentance in St. Symeon the New Theologian," *Studia Patristica* 35 (2001): 114.
49. Gregory Palamas, *Homily 34*, *PG* 151, 432D. Mark the Hermit, *On Repentance*, 1, *PG* 65, 965BC.
50. 2 Peter 3,9.
51. Mantzaridis, *Christian Ethics II*, p. 116. Symeon the New Theologian, *Ethical Discourses*, 10,11–35, ed. J. Darrouzes, *SC* 129 (Paris: 1967), p. 260.
52. Symeon the New Theologian, *Ethical Discourses*, 10,119–126, *SC* 122, p. 268.
53. Basil the Great, *Isaiah*, 119, *PG* 31, 312A. Luke 11,20. Mantzaridis, *Christian Ethics II*, p.112. John 3,2.
54. Mantzaridis, *Christian Ethics II*, p. 35.
55. John Damascene, *On Holy Images*, 1, *PG* 94, 1300AB.
56. Epistle to the Romans 8,22.

57. Konstantinos Skouteris, *Man According to the Word of the Divine Economy*.

58. Mark 9,1. Luke 9,27. Gregory Palamas, *In Favour of the Holy Resting* 1,3,26, ed. P. Christou. Gregory Palamas, *Writings*, vol. 1, pp. 436, 437.

59. Nicodemos the Hagiorite, *A Handbook of Spiritual Counsel*, ed. Panagopoulos (Athens: 2001), pp. 132, 159–160, 164. Nikolaos Zacharopoulos, "Saint Nicodemus the Hagiorite as a Teacher of Spiritual Life," *Orthodox Spirituality, Christianity-Marxism*, ed. Thessaloniki Theologians' Seminar (Thessaloniki: 1968), p. 51.

60. Nicodemos the Hagiorite, *Ethics of Christians*, p. 443.

61. Nicodemos the Hagiorite, *Ethics of Christians*, p. 440. Matthew 11,12.

62. Maximus the Confessor, *Mystagogy*, p. 142.

63. Dimitrios Tselengidis, *The Soteriology of Luther*, pp. 243, 244.

64. Luke 17,21.

65. Ezekiel 37,5–14.

66. Chrysostomos Savvatos, metropolitan bishop of Messenia, "Man and the Evolution of Modern Technology. A Challenge for Orthodox Theology," p. 15. Gregory the Theologian, *Discourse G, Epitaph to Brother Caesarion*, XXIII, *PG* 35, 785C.

67. Gregory of Nyssa, *Catechetical Discourse, Writings, GCF* vol. 1 (Thessaloniki: 1979), p. 412.

68. Dimitrios Tselengidis, *Philokalia of the Holy Neptic (Fathers)*, vol. A', ed. Holy Monastery of the Assumption of Theotokos, Boura (Arcadia: 2010), p. 19. Kallistos Katafigiotis, "The Surviving Documents on the Divine Union and Theoretical Life," *Philokalia of the Holy Neptic (Fathers)*, vol. E', (Athens: Astir, 1992), p. 23. Gregory of Sinai, *Philokalia of the Holy Neptic (Fathers)*, vol. D' (Athens: Astir, 1991), p. 52.

69. John Damascene, *Against Manichaeus*, *PG* 94, 1548A. Basil of Caesarea, *God Is Not the Author of Evil*, *PG* 31, 345A. Epistle to the Romans 6,23. Psalm 72,27. Vassilios Tsigos, *Dogmatics of the Orthodox Church* (Thessaloniki: Ostracon Publishing, 2017), pp. 241–245.

70. Gregory Palamas, *Homily 16*, *PG* 151, 189B.

71. Nicodemos the Hagiorite, *A Handbook of Spiritual Counsel*, p. 279. Katafigiotis, "The Surviving Documents on the Divine Union and Theoretical Life," p. 14.

72. Maximus the Confessor, *Letter 2*, *PG* 91, 396CD.

73. Dimitrios Tselengidis, *Philokalia of the Holy Neptic (Fathers)*, p. 19. Katafigiotis, "The Surviving Documents on the Divine Union and Theoretical Life," vol. E', p. 23.

74. Nicodemos the Hagiorite, *Eortodromion*, p. 317. Katafigiotis, "The Surviving Documents on the Divine Union and Theoretical Life," pp. 22, 23.

75. Nicodemos the Hagiorite, *Eortodromion*, vol. B', p. 283. Matthew 5,8.

76. Nicodemos the Hagiorite, *A Handbook of Spiritual Counsel*, p. 283. Kallistos, Patriarch, "Chapters on Prayer," *Philokalia of the Holy Neptic (Fathers)*, vol. D' (Athens: Astir, 1991), pp. 331–333, 365. Katafigiotis, "The Surviving Documents on the Divine Union and Theoretical Life," pp. 6, 7. Citterio Elia, *Nicodemo Agiorita, in La Theologie Byzantine et sa Tradition II* (Turnout: Brepols Publishers, 2002), pp. 961–965

77. Nicodemos the Hagiorite, *New Klimax*, pp. 110, 111. Citterio Elia, *Nicodemo Agiorita*, pp. 961–965. Peter of Damascus, "On the Seven Bodily Acts," *Philokalia of the Holy Neptic (Fathers)*, vol. C' (Athens: Astir, 1991), p. 25.

78. Nicodemos the Hagiorite, *New Klimax*, p. 163. Katafigiotis, "The Surviving Documents on the Divine Union and Theoretical Life," pp. 12–17. Matthew 5,8.

79. Nicodemos the Hagiorite, *A Handbook of Spiritual Counsel*, p. 284. Katafigiotis, "The Surviving Documents on the Divine Union and Theoretical Life," pp. 50, 54.

80. Nicodemos the Hagiorite, *A Handbook of Spiritual Counsel*, pp. 273–274. Kallistos, Patriarch, "Chapters on Prayer," pp. 338, 339.

81. Nicodemos the Hagiorite, *A Handbook of Spiritual Counsel*, p. 276. Katafigiotis, "The Surviving Documents on the Divine Union and Theoretical Life," p. 12. Theoleptos of Philadelphia, *Philokalia of the Holy Neptic (Fathers)*, p. 10.

82. Kallistos, Patriarch, "Chapters on Prayer," pp. 358, 359. Kallistos Katafigiotis, "The Surviving Documents on the Divine Union and Theoretical Life," pp. 54, 55.

83. Symeon the New Theologian, *The Ethical Discourses* vol. 1, a', 123–127, *SC 122*, p. 182.

84. Symeon the New Theologian, Chapters on the Holy Trinity, 8, *PG* 77, 1232D. *The Ethical Discourses* vol. 1, a', 127–133, *SC 122*, p. 182.

# Conclusion

Artificial intelligence is part of the broader picture of the achievements of technological progress with unique potential on a global level. The applications resulting from the use of AI affect a large proportion of human activities with many positive results so far.

AI significantly impacts the economy, politics, culture, information, governance, health, security, transport, and communication. With the rapid development of AI technology, new possibilities and perspectives are emerging, but at the same time, new challenges arise from its proper use.

It is necessary to have a proper understanding of what AI is, how it relates to technology, how it is connected to other sciences, and how this association of AI with a range of sciences and applications is accessible to the life of modern humanity and society. A proper understanding can rid theories of phobic syndromes or extreme approaches regarding its use and applications in the present and the future.

As a result of advancements and processes, apart from the purely technocratic part of the evolution of AI, the concept of "human-centered AI" is developed to set the balancing framework for the use of AI.

Human-centered AI and ethical AI set the anthropocentric orientation of AI development and evolution so that AI applications' progress holds humanity as its main reference point. The interaction of humans with AI is a given, and in the context of its applications, it is important to safeguard human rights and freedoms.

Yet, the anthropocentric ethics of AI can be interpreted in many ways.

The fluidity of the interpretations and boundaries of ethics emerging from the functionality of AI has highlighted the need for a defined ethical code. In determining the ethical principles for programming and monitoring artificial intelligent entities, ethical criteria are set using "ethical algorithms" to provide an ethical system that, based on human rights, will impact practical applications, particularly in cases where AI systems are capable of autonomous decision-making. In recent years, governments and international organizations have drawn up ethical codes of conduct in line with human values and rights to safeguard citizens' political, economic, social, and cultural rights. For them to be considered reliable, AI applications must meet the criteria of human intervention and oversight, transparency, accountability, social and environmental well-being, technical excellence and safety, diversity, justice and governance, and non-discrimination.

The use of AI in governance has produced excellent results without excluding cases of abuse and violation of human rights and freedoms. Totalitarianism can be reinforced using AI, and political, economic, and military power rivalries can be triggered with information and its freedom. However, apart from the risks, there is also an upside to the proper use of AI in governance.

There is excitement over the use and development of AI-led individuals or trends to develop a new anthropology and theology influenced by Protestant or Chiliasm theories with an apocalyptic and messianic orientation. Regarding the newly ventured anthropology, the notion of "soul" was attributed to technocratic realities and digital entities, as well as the concepts of deity or God, religion, or the Church. At the same time, the role of AI is linked to existing religions by its contribution to the understanding of spiritual content and its rendering by algorithms through the applications of AI. The question is: no matter how clever an AI system is, can it comprehend the spiritual teachings of religions and render them?

## IS IT POSSIBLE TO ATTRIBUTE DEITY, GOD, RELIGION, OR THE CHURCH TO HUMAN ACHIEVEMENT?

The "apocalyptic AI" enthusiasts proclaim a transcendent Virtual Kingdom in cyberspace, where humans will upload their minds into machines and enjoy a virtual reality paradise with perfect virtual bodies. They proclaim unique immortality and salvation by uploading human intelligence and emotions into digital entities within the emerging and forthcoming Virtual Kingdom. The soul takes on the concept of a digital entity and the body the idea of a virtual one so that humans can achieve immortality by "uploading" their minds into AI systems. The personality of a human can be "uploaded" and "saved" by creating a digital entity in a Virtual Kingdom.

This proclaimed immortality ensures the fulfillment of transhuman salvation in a non-physical Virtual Kingdom of cyberspace.

These approaches led to the creation of the religion or Church of Transhumanists, attributing to AI the term deity with purely cognitive and technocratic characteristics. The main argument for this deification of AI is that super-intelligence will have greater perceptual capacity and knowledge than humans and, therefore, could be trusted by them. The cognitive nature of this project is apparent by the fact that an attempt is being made to create a super-religion, which will be the result of the synthesis of the teachings of existing religions based on knowledge and composition to shape new teaching for the benefit of all people and the "improvement of society." Thus, this is a new form of Gnosticism: the technocratic, digital, or algorithmic Gnosticism.

There is also a creeping objective, Gnosticism, according to which a living person's value stems from their knowledge. Therefore, an algorithmic entity with increased knowledge would be of greater value than humanity, overlooking the spiritual dimension of humankind as created in the image and likeness of God. The issue is further exacerbated when an artificial entity or algorithm, based on its cognitive value, makes autonomous decisions for humanity.

Applying this new type of Gnosticism can also occur within existing religions through the devotional uses of AI applications. For example, there are applications designed for confession or prayer. In this context, a legitimate question arises: Can an AI application perform the sacrament of confession? The approach would be more of a dry, emotionless, logocratic, and gnostic character than a spiritual one. Similarly, the question about prayer: Can an AI application pray for humanity?

The approaches mentioned above meet with reactions ranging from reservations to outright criticisms and responses regarding turning AI into a deity, religion, or Church, but also for its use in devotional applications.

With the rapid development of modern technology, mental capacity becomes the main and exclusive factor of a "new" identity for humanity, and the creation of thinking entities is reinforced, i.e., digital technocratic systems that do not require a human body to generate cognition, overlooking the psychosomatic unity of humanity.

In this context, the critical consideration of such proposals is beneficial for those guided by principles and values, such as those described in the second part of this work. Indeed, the holistic identity approach to humanity is not determined by a unilateral development of the psyche or intelligence or by an absolutized enforcement of intellectual capacity at the expense of biological and physical constitution, and even in contrast to them. A one-sided understanding of human nature, which glorifies intellectual faculties by downgrading the rest of human nature, runs the risk of devaluing and degrading the human body, even as an instrument of cognition and knowledge. Moreover, digital cognition, considered to be "evolved" as described above, with the help of AI, cannot replace or evolve feelings and experiences, such as love, which, together with cognition, have a spiritual orientation.

Furthermore, in addition to humanity's freedom of thought, self-authority, reason, and the whole creation of humankind as the image and likeness of

God, which received life from the breath of God, cannot be replaced by algorithmic schemes. Similarly, the proclaimed bliss of the Virtual Kingdom cannot substitute the divine bliss that arises from humanity's communion with God.

The Orthodox Church, based on the patristic tradition and Christian ethics, also points out the dangers latent in certain scientific achievements when these, from operational instruments, are absolutized and attributed ontological properties, as attempted by the absolutized proposals of AI's deification enthusiasts. In contradiction to these deviations, Christian ethics gives prominence to humanity's dignity and divine destiny. At the same time, it respects the freedom of scientific research and identifies its positive contribution to humankind, criticizing the misuse of science associated with ideologies dangerous to humanity.

From the above, it can be concluded that for the adequate investigation of the moral issues raised and, more generally, the evaluation of the use of AI and its purposes, another approach is necessary, the criteria of which are responsive to the proposals of the transhumanists on the religion and theology of AI. In this respect, developing a theological approach to humanity becomes essential. For this very reason, the second part of this study explores the criteria and principles of the approach of artificial intelligence to humanity, which will truly benefit humanity. The set criteria of Christian ethical teaching and the values derived from them are the shield against any ideology directed against promoting human totality as an individual and a society. Their mere reference is sufficient for the understanding of inherently healthy practices. For this reason, in the present work, we have chosen to quote them without any further attempt to argue or contest them.

As has already been established, the reasoning is moral-anthropological and religious-theological. However, to adequately examine this crucial issue, it is imperative to develop a theological approach to humanity according to the relevant teachings of the Eastern Orthodox Church.

The dominant role of humanity in creation is directly related to the soteriological perspective of their creation, based on the image and likeness, as human nature has an ontological referentiality towards God, and humanity's existence has a theocentric character and direction. Humanity was created in the image of God to be like God.

Humanity, as a psychosomatic unity, was created by God as the pinnacle of divine creation. God created humanity as a dual being, made of two different natures and essences, from creation's invisible and visible elements.

The matter refers to the physical world, and the soul to the spiritual one. The nature of the body links humanity to the tangible and material, and the soul to the mental and heavenly, so that humans can commune with the supernal, the mental, and the heavenly. Humanity is a composite of soul and body; the body draws life and strength from the soul, while the physical state affects the mental state.

The mental logical power relates to the emotional part, the desiring part, with the senses, which are inherent and interact with the body. The mind and soul permeate human existence as an invisible and mental nature. When the rational part prevails over the emotional and the desiring, through the proper management of the senses, the rational prevails over the irrational. It is willingly led by virtue to perfection. At the same time, through the appropriate use of self-authority, humans are healed by renewing grace so that they can become children of God by grace and holy like God the Father.

Despite the ontological gap between created humanity and the uncreated God, the uncreated grace of God transforms the whole person into a godlike one, and created humanity is united with God as a single psychosomatic existence and enters a supernatural state of theosis through the illumination of divine light, as in the event of the Transfiguration of the Lord.

After the Fall, and according to the plan of the Divine Economy, the Creator God renewed humanity through the incarnation of the Son and Word of God. Through Jesus Christ in the Church, humanity is regenerated as a "new human." Through the incarnation of the Son and Word of God and the assumption of the whole person, the God-man Jesus Christ healed corruption, reshaping and orienting the entire person towards the "likeness." Thus, humanity enters a second communion with God, superior to the first (the pre-Fall one), which results in the theosis of the assumed human nature by God the Word. The "new human" is built within the Church, where through the holy sacraments they put on Christ, are renewed, and partake sacramentally of His death and Resurrection, and can be perfected "into a perfect human, to the measure of the age of the fullness of Christ."

The renewed human becomes a communicant of the divine nature, the mystical body of Christ, through the sacraments of the Church. The One, Holy, Catholic, and Apostolic Church as the body of Christ in the image of the Holy Trinity is a God-human communion, the communion of the Holy Trinity with the new human. The Church has a Trinitarian character, as revealed in the New Testament, as the people of God the Father, the body of Christ, and the temple and communion of the Holy Spirit. Everything in the Church derives from the Father through the Son in the Holy Spirit. In the context of the Divine Economy, where one person of the Holy Trinity acts, the other persons are also present, declaring the unity of the Holy Trinity.

The Church is the image of the Triune God, within which the union of all things and the likeness of God is accomplished. Within the Church, the believer becomes God by grace so that theosis has an ecclesiological character. Within the Church, the liturgical unity of the world and humanity is achieved through uncreated actions which act sanctifyingly and unifyingly as the communion of humanity and creation with God. The Church has a charismatic character with an eschatological perspective.

The Christocentric character of the Church is also determined by the direct relation of the Church to the Holy Sacraments, which, as "doors," provide God's uncreated grace. Through spiritual therapy by the Sacraments, the believer receives spiritual strength and is healed, purified, and sanctified in soul and body, the whole person. The sacramental life for every member of the Church has a salvific perspective.

Yet the sacramental life is linked to the application of the commandments and the practice of virtue. Virtue has a Christocentric and ecclesiastical character with an eschatological perspective. The doctrinal teaching of the Church is consistent with the believers' life in Christ on their path to perfection, reflecting the ethical standards of the believer's life.

Orthodox Christian ethics has direct references and relationships with Orthodox theology and patristic literature. Ethics is governed by humanity's relationship with God, fellow humans, and creation. Thus, ethics has a theocentric character, with the relationship between humanity and God as its main reference point. It determines human behavior based on the code of values, which, in turn, is defined by the virtues. The standard of virtues is Christ Himself as the norm of moral conduct and sociality.

Christ is the standard for humanity's morality, as the believer, by imitating Christ's morality, partakes in the sacrifice of the cross and the expectation of the resurrection within the body of the Church. Moreover, the Beatitudes define Christian ethics.

Given that God created humanity in His image and potentially in His likeness, the Beatitudes define Christian morality in conjunction with the imitation of Christ. The imitation of Christ concludes with the pinnacle of virtues, the love towards God and humans, and healthy self-love, for love unites everyone and everything in a bond of peace. Love and peace have a theocentric character because their source is God, and the love between God and humanity also defines the relations of love between humans.

The cultivation of virtues and participation in the Sacraments of the Church lead humanity to salvation as a gift from all three persons of the Holy Trinity. The salvation of humanity is the work of the Triune God, and when one person of the Holy Trinity acts, the whole Trinity works together. The salvific work of the incarnation of the Son and the Word is a work of the Father's love and favor and the Holy Spirit's energy and grace.

According to the soteriology of Athanasius the Great, the acceptance of the fact of the incarnation of the Son and Word of God, that is, faith in Jesus Christ, constitutes the soteriological basis of humanity's synergy with the divine grace which leads to theosis.

The salvation of humanity is realized within the Church, the sacramental body of Christ, where the Sacraments are performed with the renewing grace as the joint sanctifying energy of the Holy Trinity. Thus, salvation is accomplished by cultivating virtues, which leads to apathy and is completed by the synergy of the Triune God.

Salvation within the Church has an eschatological character since it orients the believer towards the eschata, the gift of the heritage of the Kingdom of God. The Kingdom of God was revealed with the coming of Christ in the world and history. It is happening in the present, but in its entirety, it will become apparent in the age to come, as the eschaton is experienced in anticipation as the present; therefore, the present has also assumed an eschatological character.

Christ opened the way for the resurrection of the believers and the path to heaven, and within His sacramental body, the Church, through the sacraments, His glory and His final Kingdom are revealed.

The Revelation of God the Father through Jesus Christ in the Holy Spirit has a praising character with its main attribute the Hope of Christ's coming to the eschata in Divine glory.

On the last eighth day, the unfading day of the Lord, Christ will be revealed anew in Glory, and the saved will behold the Glory of God in the eternal joy of the Kingdom of God. However, the mercy of God merits the Saints to foretaste His eternal glory and to have revealed to them, through sight, the unspoken mysteries of the last days of His future Kingdom.

In summary, by examining the theological approaches to the relationship between humanity and artificial intelligence, it is observed that when humanity's attitude towards life and things is adopted without any reference to the cause of everything, which is God, controversial ideologies and practices emerge, culminating in materialism, ascribing greater value to the created-tangible reality and absolutizing the material goods. This results either in humanity ending up in self-theosis with the characteristics of egoism of the Fall or in the deification of creation and its achievements, with the attributes of idolatry.

The Orthodox Christian view of things and life actualizes the transition from perishable to imperishable, from created to uncreated, and from tangible to mental. The divine energy is the cause of beings; it derives from the existing God, to whom all beings owe their existence and their remaining existence.

The essence of God cannot be communicated from its creations. The relationship of the design to the Triune God is interpreted and understood in the context of the distinction between the communed actions and the uncommuned and unapproachable divine essence. Since God's essence cannot be communicated, God's communion relationship with the creation depends on His salvific and creative energy. Thus, the communion of the Triune God with creation and humanity is an energetic relationship with soteriological implications in the context of the Divine Economy.

Although humanity has gone through phases of revolutionary technological change in the past, despite deviations, over time, a new equilibrium has prevailed. The predominant factor of equilibrium, in the past and present, is the identification of the concept of ethics and its application in the use of technology.

The ethics of AI promotes social eudaimonism, as the good and the useful are defined through a material-utilitarian reference to the social whole and not to individuals and their wholesome moral being.

The development of artificial intelligence is the narrative of a universal perfection (technological) of humanity, with a disproportionate emphasis placed on personal moral perfection and embellished with superficial philosophical criteria and not with the genuinely wholesome moral-theological-anthropological criteria that promote the wholeness of humanity. In this way, the existence of artificial persons emerges, which may seriously undermine the very psychosomatic nature of humanity. At the same time, it tries to base human value on an intrinsic, impersonal, undefined, and independent from God nature, which it ultimately deifies.

Questions also arise from other proposals of the transhumanists, such as if it is indeed possible for AI to reach the point of exceeding human intelligence, can subsequent events be predicted?

If humanity were to allow AI researchers to create autonomous intelligent machines, would safeguards against unethical behavior be put in place?

At the same time, apart from the possibility of AI simulating human thought, there is the question of whether achieving a combined simulation of the emotions of love, hope, and joy on a conscious level is feasible.

Ethics is connected to shaping humanity's conscience, which is also determined by the cultivation provided by the environment and the culture one receives. On a moral level, conscience is defined solely by the natural or cosmic law or the spiritual and eternal. In the latter case, as Christian ethics, it follows the example of Jesus Christ. Thus, while conscience has a universal reference to humanity, it is shaped by each person's acceptance of either a unilateral natural morality or a spiritual approach and perspective.

This dialectic on moral consciousness also underlies the reflection on the evaluation and utilization of AI. The ongoing debate on AI ethics extends to assessing the impact of these technologies on human rights so that AI can operate on a level where its self-autonomy does not compromise human freedom and rights.

Human rights are directly connected with the commandments of God and their observance, in conjunction with the application of justice in the everyday life of humans. The respect for human rights is related to their uniqueness as created "in the image and likeness" of God.

Orthodox Christian ethics is experienced and forged within the Church, centered on the person and the circumstances of the admission of their value and with love as a basic component. As the body of Christ, the Church is also composed of its faithful members, functioning using a life of worship towards its head and a mystical dimension to sponsor the benefits of divine energy (divine grace).

According to the "apocalyptic AI" enthusiasts, AI proclaims "worldly immortality" and the creation of super-intelligent worldly entities. Orthodox spirituality is based on a perspective that transcends the present tangible world, on the salvific energy of the Triune God that arises through the experiential approach of the Church's doctrines through the inner and sacramental life.

The revelation of God to humanity is reflected in Scripture to reveal His creative, providential energy as much as possible and educate with the perspective of the salvific dimension.

The "apocalyptic" Orthodox theology is of a purely ecclesiological, Christocentric, and trinitarian character. The revelation of God in the past, present, and future is within the framework of Divine Economy for the salvation of humanity as a result of God's charity and love for humanity. As a Messiah, Christ revealed Himself to the world as a Mediator and Savior between God and humanity.

In contrast, apocalyptic AI envisions a Virtual Kingdom as a creation of humanity and technology. The Virtual Kingdom will be enjoyed by digital entities, both autonomous and uploaded human consciousnesses. These algorithmic consciousnesses cannot replace the human soul or body through a virtual body. Consequently, apocalyptic AI proclaims a technological "autotheosis" of humanity with their salvation trapped in a Virtual Kingdom of algorithms.

The salvation of humanity, according to the apocalyptic AI, transhumanism, Terasem, Turing, and Singularity supporters, involves the creation of a technological transhuman who will be saved in the Virtual Kingdoms without the burden of bodies. But is salvation possible by a human creation? Is salvation possible without the practice and the journey towards moral perfection? Is salvation possible without participation in the Sacraments of the Church? Is salvation possible without faith in Jesus Christ? Is salvation possible outside the Church of Christ?

How can a human creation be qualified as a deity? Is it possible for technology to assume the character of a religion? Is it possible for there to be a Church without Christ?

Could transhumanism resemble the egotistic enthusiasm of the Tower of Babel? With the difference, of course, that while in Babel, humanity was trying to reach heaven; with the "deification" of AI, an autonomous technological algorithmic deity knows, foresees, and intrudes into the world and humanity from the Virtual Kingdom.

Science and technology are gifts from God to humanity. However, to use technology properly, the moral cultivation of humanity is necessary. Moral cultivation is often perceived beyond the theocentric, Christocentric, and ecclesiological dimensions, resulting in a purely secular and stale anthropocentric character. Humanity's moral perfection can improve relations with the environment and fellow humans when defined by the apostolic and patristic experience of the Church as Christ taught it.

The misunderstanding of morality and, consequently, spirituality leads to abuses and misinterpretations regarding the use and value of technology, with its extreme and dangerous applications. Moreover, the misinterpretations lead to erroneous analogies of technological discovery and its achievements by giving it the title of deity, with purely scientific or cognitive characteristics, or even the misuse of the term "Church."

In conclusion, based on its theology and tradition, the Orthodox Church provides sound anthropological and theological criteria for properly using technology, specifically AI, centered on humanity's self-consciousness, which stems from their relationship with God.

Reflecting on Philip Dick's inquiry: "Do Androids Dream of Electric Sheep," we arrive at an equally interesting question: "What religion will AI robots have?"

# Bibliography

Aatif, Sulleyman.. "Elon Musk Slams Proposal to Create an Artificial Intelligence 'God' That People Will Worship." *Independent*, 2017. https://www.independent.co.uk/life-style/gadgets-and-tech/news/ai-god-elon-musk-artificial-intelligence-religion-anthony-levandowski-way-future-a8017296.html (12/12/2020).

Agouridis, Savvas. *John the Evangelist, Introductory, Explanatory and Theological Studies on the Fourth Gospel.* Athens: University of Athens, 1975.

Alexiadou, Anastasia-Sofia.. "Artificial Intelligence and the COVID-19 Pandemic. Moral Implications." *Archives of Hellenic Medicine* 37, no. 5 (2020): 708–713.

Amodei, Dario, Chris Olah, Jacob Steinhardt, Paul Christiano, John Schulman, and Dan Mané. *Concrete Problems in AI Safety.* June 21, 2016. http://arxiv.org/abs/1606.06565 (07/12/2020).

Anthimos, Gazis. *Interpretation of Genesis*, *PG* 87, 128C.

Aquinas, Thomas. *Summa Theologica* I, Question 93, Article 6. https://www.documentacatholicaomnia.eu/03d/12251274,_Thomas_Aquinas,_Summa_Theologiae_%5B1%5D,_EN.pdf (01/02/2021).

Aristotle. *Politics* 3,9 (14),4. 7,2.

Armstrong, Stuart, Nick Bostrom, and Carl Shulman.. "Racing to the Precipice: A Model of Artificial Intelligence Development." *Technical Report. Future of Humanity Institute,* 2013-1, pp. 1–8. https://www.fhi.ox.ac.uk/wp-content/uploads/Racing-to-the-precipice-a-model-of-artificial-intelligence-development.pdf (09/12/2020).

Athanasius the Great, *Against thc Arians*, Discoursc 1, 16 *PG* 26, 45BC. 2, 41, *PG* 26, 236A. 1, 39 *PG* 26, 92C. 2,70, *PG* 26, 296. 2, 65, *PG* 26,3.

———. *Against the Greeks*, 41, *PG* 25, 84A.

———. *Discourse on the Incarnation of the Word and through the Body to Ourselves*, *PG* 25, 95–198.

———. *Discourse on the Incarnation of the Word*, 17, *PG* 25, 125AB. 5, *PG* 20,101B.

———. *On the Incarnation Discourse against the Arians*, 12, *PG* 26, 1004B.

———. *To Serapion* 1, 28, *PG* 26, 596ABC.

———. *To the Antiochian Lord*, *PG* 28, 608A.

Athenagoras. *On the Resurrection of the Dead*, *PG* 6, 1000B.

Babiniotis, Georgios. *Dictionary of Modern Greek.* Athens: Lexicology Center Ltd, 1998.

Bassil-Morozow, Helena.. "God Is an Algorithm: Why We're Closer to a Black Mirror-Style Reality than We Think." *The Conversation*, January 31 2018. https://theconversation.com/god-is-an-algorithm-why-were-closer-to-a-black-mirror-style-reality-than-we-think-90669 (01/02/2021).

Basil of Caesarea. *God Is Not the Author of Evil*, *PG* 31, 345A.

———. *Homily C', On the Care of Thee*, *PG* 31, 215.

———. *On the Holy Spirit*, *PG* 32,128–192C.

———. *Ps.* 32,6, *PG* 29, 337.

Basil the Great. *Isaiah 1*, 13, *PG* 30,140–312A.

———. *On the Care of Thee*, *PG* 31, 212B.

———. *Conditions by Amplitude*, 43, *PG* 31, 1028BC.

Benedikt, Michael. *Introduction. In Cyberspace: First Steps*. Cambridge, MA: M.I.T. Press, 1994.

Benek, Christopher. "Thinking West: Robots Taking Selfies." *El Portal Literary Journal*, February 25, 2015. https://elportaljournal.com/category/weird-s//(05/03/2021).

Bizzaco, Michael, Erika Rawes, and Kim Wetzel.. "What Is Amazon's Alexa, and What Can It Do?" March 30, 2021. https://www.digitaltrends.com/home/what-is-amazons-alexa-and-what-can-it-do/ (10/04/2021.

Bostrom, Nick. *Superintelligence: Paths, Dangers, Strategies*. Oxford: Oxford University Press 2014.

Bostrom, Nick, Allan Dafoe, and Flynn Carrick.. "Public Policy and Superintelligent AI: A Vector Field Approach." *Future of Humanity Institute*, 2018. http://www.nickbostrom.com/papers/aipolicy.pdf (03/12/2020).

Bostrom, Nick, and Eliezer Yudkowsky. *The Ethics of Artificial Intelligenc*e, ed. In *The Cambridge Handbook of Artificial Intelligence*, edited by K. Frankish, W. M. Ramsey, 316–334. Cambridge: Cambridge University Press, 2011. https://www.fhi.ox.ac.uk/bostrom-n-the-ethics-of-artificial-intelligence%C2%94-in-the-cambridge-handbook-of-artificial-intelligence-eds-william-ramsey-and-keith-frankish-cambridge-cambridge-university-press-2014-w/ (11/11/2020).

Braciotos, Nikolaos. *Anthropology of the Old Testament. The Man as Divine Creation*. Athens: 1989.

Brantlinger, Patrick. "The Gothic Origins of Science Fiction." *Novel: A Forum on Fiction* 14, no. 1 (Autumn 1980): 30–43.

Brooks, Rodney Allen. *Flesh and Machine Robots Will Change Us*. New York: Pantheon 2002.

Brown, Dan. *Origin: A Novel*. New York: Doubleday 2017.

Carlos, Ian. "Ex-Google Engineer Anthony Levandowski Has Closed His Artificial Intelligence Church." *The Verge*, February 19, 2021. https: //www.theverge.com/2021/2/19/222291769/anthony-levandowski-church-closed-artificial-intelligence-waymo-uber/ (11/02/2021).

CBS News. *The Evolution of Dan Brown*. 2017. https: //www.cbsnews.com/video/theevolution-of-dan-brown/ (19/01/2021).

Ceif, George. *Deep Learning vs Classical Machine Learning*. 2018. https://towardsdatascience.com/deep-learning-vs-classical-machine-learning-9a42c6d48aa (10/12/2020.

Chinadaily. "Governance Principles for the New Generation of Artificial Intelligence: Develop Responsible Artificial Intelligence." 2019, n 22, Principle 1. http://www.chinadaily.com.cn/a/201906/17/WS5d07486ba3103dbf14328ab7.html (04/02/2021).

Chinese National Governance Committee for the New Generation Artificial Intelligence, led by China's Ministry of Science and Technology. "Governance Principles for a New Generation of Artificial Intelligence: Develop Responsible Artificial Intelligence." 2019, n 22, Principle 1. http://www.chinadaily.com.cn/a/201906/17/ WS5d07486ba3103dbf14328ab7.html (04/02/2021).

Christou, Panagiotis. *Gregory Palamas' Writings*, vol. A', Thessaloniki: 1962.

———. *The Teachings on Man of Theophilus of Antioch.* Thessaloniki: 1957.

Chymebot. *Confessions of a Chatbot.* https://chymebot. com/blog/confessions-of-a-chatbot/ (12/012/2020).

Clement of Alexandria. *Paedagogus* 1,6, *PG* 8, 21–388.

———. *Prophetic Utterances*, 22, *PG* 9, 708.

———. *Stromata*, 6,15, *PG* 9, 516B,13.

Coeckelbergh, Mark. "Virtual Moral Agency, Virtual Moral Responsibility: On the Moral Significance of the Appearance, Perception, and Performance of Artificial Agents." *Springer, AI & Society, Journal of Knowledge, Culture and Communication* 24 (2009): 181–189. https://link.springer. com/article/10.1007%2Fs00146-009-0208-3 (11/11/2020).

Collins, Joseph John. "From Prophecy to Apocalypticism: The Expectation of the End." In *The Encyclopedia of Apocalypticism* Volume I: *The Origins of Apocalypticism in Judaism and Christianity,* edited by John J. Collins, 129–161. New York: Continuum Press, 2000b.

———. *The Apocalyptic Imagination: An Introduction to the Jewish Matrix of Christianity.* Crossroad, NY: 1984.

Corncll, Jay, R.U. Sirius. *Transcendence. The Disinformation Encyclopedia of Transhumanism and the Singularity Paperback.* New York: Disinformation Company Ltd, 2015.

Council of Europe, European Commission for the Efficiency of Justice. "European Ethical Charter on the Use of Artificial Intelligence in Judicial Systems and Their Environment." 2018 Principle 1. https://rm.coe.int/ethical-charter-en-for-publication-4-december-2018/16808f699c (03/02/2021).

Cramer, Anthony. *Catenae graecorum patrum in Novum Testamentum*. University of Michigan Library, 2009.

Cramer, Maria. "A.I. Drone May Have Acted on Its Own in Attacking Fighters, U.N. Says." *New York Times,* June 3, 2021. https://www.nytimes.com/2021/06/03/world/africa/libya-drone.html (17/07/2021).

Cyprian of Carthage. *Epistle*, 73,21, *PL* 3, 1169A.

Cyril of Alexandria. *On First Corinthians*, 12,7, *PG* 74, 885.

———. *Commentary on John*, *PG* 74, 560–700.

———. *Commentary on Isaiah, PG* 70, 236B–1217A.
———. *Commentary on John*, 7-8, *PG* 74, 92D–433B.
———. *Commentary on John*, 2,1, *PG* 73,228D.
———. *Commentary on John*, 9, 1, *PG* 74, 280A.
———. *Commentary on Luke*, *PG* 72, 485–909A.
———. *On Matthew PG* 72, 456C.
———. *Festive Homilies, Homily* 10, 1, *PG* 77, 497A–609C.
———. *On the Holy and Consubstantial Trinity*, A', 3, *PG* 75, 853C.
———. *On the Incarnation* 3,28, *PG* 75, 1452B–C /1468B–C.
———. *On the Veneration and Worship in Spirit and Truth* B΄, *PG* 68, 238D–1117A.
———. *On the Holy Trinity*, 1,3, *PG* 75, 808–1044B.
———. *Address to the Wueens* 2, 46, *PG* 76, 400B–C.
———. *Commentary on John*, *PG* 73, 65A.
———. *Commentary on John*, *PG* 74, 557D.
———. *Commentary on Peter*, 2, 6, 7, *PG* 74, 1013A.
Cyril of Jerusalem. *Mystagogical Catecheses*, 2,5, *PG* 33, 1081A.
Dafoe, Allan. "AI Governance: A Research Agenda." *Governance of AI Program Future of Humanity Institute.* First Draft July 2017; August 27, 2018. https://www.fhi.ox.ac.uk/wp-content/uploads/GovAIAgenda.pdf (03/12/2020).
Daniélou, Jean. *Platonisme et théologie mystique. Essai sur la doctrine spirituelle de Saint Grégoire de Nysse.* Paris: 1944.
Davincisurgery. "Intuitive for Patients, Robotic-Assisted Surgery as a Minimally Invasive Option." www.davincisurgery.com (14/01/ 2021).
De Garis, Hugo. *The Artilect War: Cosmists vs. Terrans: A Bitter Controversy Concerning Whether Humanity Should Build Godlike Massively Intelligent Machines*. Palm Springs: ETC Publications, 2005.
Delikostopoulos, Athanasios. *Greek Thought and Christian Conscience*. Apostoliki Diakonia, 1993. http: //www.apostoliki diakonia.gr/gr_main/catehism/theologia_zoi/themata.asp?contents=ecclesia_history/contents_texts.asp&main=texts&file=4.htm (15/02/20221).
Dembski, William. "Conflating Matter and Mind." *Perspectives on Science and Christian Faith* 43, no. 2 (June 1991): 107–111.
Descartes, Rene. *Discourse on the Method of Rightly Conducting the Reason, and Seeking Truth in the Sciences* https://www. gutenberg.org/files/59/59-h/59-h.htm (10/01/2021.
Dignum, Virginia. "Ethics in Artificial Intelligence: Introduction to the Special Issue. Ethics and Information Technology." *Springer, Ethics and Information Technology* 20 (2018): 1–3. https://link.springer.com/article/10.1007/ s10676-018-9450-z (11/11/2020).
Dimitropoulos, Panagiotis. *The Anthropology of Athanasius the Great.* Athens: 1954.
———. *Orthodox Christian Ethics.* Athens: 1970.
Dionysius the Areopagite. *Epistle 10*, *PG* 3, 1117A–1117B.
———. *On Divine Names*, 4,15, *PG* 3,713A.
———. *On the Ecclesiastical Hierarchy* 4, *PG* 3, 376A.
Einstein, Albert. *Out of My Later Years.* New York: Philosophical Library, 1950.

Elia, Citterio. *Nicodemo Agiorita, in La Theologie Byzantine et sa Tradition II.* Brepols Publishers, Turnout 2002.

Ellul, Jacques. *The Technological Society.* New York: Vintage Books, 1964.

Eppling, Christopher. *A Study of the Patristic Doctrine of Free Will.* Wake Forest, NC: 2009.

Escribano, Alberca. "Die Entdeckung des inneren Menschen." *to Gregor von Nyssa und die Philosophie, II Internationales Kolloquium über Gregor von Nyss*a, 43–58, Freckenhost bei Münster 1972. Leiden: E. J. Brill, 1976.

European Commission's High-Level Expert Group on Artificial Intelligence. "Ethics Guidelines for Trustworthy AI." 143 (2019). https://ec.europa.eu/futurium/en/node/6945#_Toc414754698 (12/12/2020).

European Commission, Communication from the Commission to the European Parliament, the Council, the European Economic and Social Committee and the Committee of the Regions. *Building Trust in Human-Centric Artificial Intelligence.* August 4, 2019, Brussels. https://ec.europa.eu/transparency/documents-register/detail?ref=COM(2019)168&lang=el(15/12/2020).

European Parliament 2019–2024. "Artificial Intelligence: Questions of Interpretation and Application of International Law." Texts adopted, Provisional version, P9_TAPROV (2021) 0009. https://www.europarl.europa.eu/doceo/document/TA-9-2021- 0009_EL.pdf (12/12/2020).

European Parliament 2019–2024. "Texts Adopted, Artificial Intelligence: Questions of Interpretation and Application of International Law, European Parliament Resolution of 20 January 2021 on Artificial Intelligence: Questions of Interpretation and Application of International Law in So Far as the EU Is Affected in the Areas of Civil and Military Uses and of State Authority Outside the Scope of Criminal Justice." 2020. https://www.europarl.europa.eu/doceo/document/TA-9-2021-0009_EL.pdf (09/01/2021).

European Union. "Regulation (EU) 2018/1725 of the European Parliament and of the Council." 2018. https://ec.europa.eu/info/law/law-topic/data-protection/data-protection-eu_el 2016-2018, https://eur-lex.europa.eu/legal-content/EN/TXT/HTML/?uri=CELEX:32018R1725&from=EN (10/03/2021).

Everitt Tom, Victoria Krakovna, Laurent Orseau, and Shane Legg. "Reinforcement Learning with a Corrupted Reward Channel." Paper presented at the 2017 International Joint Conference on Artificial Intelligence (IJCAI), 2017, 4705–4713.

Federal Trade Commission. "Big Data: A Tool for Inclusion or Exclusion? Understanding the Issues." January 2016, p. 1, 2. https:// www.ftc.gov/system/files/documents/reports/big-data-tool-inclusion-or-exclusion-understanding-issues /160106big-data-rpt.pdf (14/11/ 2020).

Feidas, Vlasios, *Ecclesiastical History*, vol. A'. Athens: 1997.

———. *Ecclesiology between Christology and Pneumatology in the Light of the Patristic Tradition.* Athens: 2018.

Filiotis-Vlahavas, Christos Protopresbyter, *The Teaching of Athanasius the Great on the Holy Spirit.* Thessaloniki: 2004.

Fjeld, Jessica, Nele Achten, Hannah Hilligoss, Adam Nagy, and Madhulika Srikumar. "Principled Artificial Intelligence: Mapping Consensus in Ethical and Rights-Based Approaches to Principles for AI." Berkman Klein Center Research Publication (2020–2021). https://dash. harvard.edu/bitstream/handle/1/42160420/HLS%20White%20Paper%20Final_v3.pdf (02/01/2021).

Future of Life. "The Asilomar AI Principles." California, January 2017. https://futureoflife.org/ai-principles (03/12/2020).

Future of Life Institute. "Asilomar AI Principles." 2017 n. 89. https://futureoflife.org/ai-principles/?cn-reloaded=1 (03/02/2021).

Georgouli, Katerina. "Artificial Intelligence, an Introductory Approach." Hellenic AcademicEbooks, Zografou 2015. http://repfiles.kallipos.gr/html_books/93/01amain.html#_idTextAnchor002,file:///C:/User/User/Downloads/%CE%A4%CE%B5%CF%87%CE%BD%CE%B7%CF%84%CE%AE%20%CE%9DCE%BF%CE%B7%CE%BC%CE%BF%CF%83%CF%8D%CE%BD%CE%B7%20(4).pdf (14/12/2020).

Geraci, Robert. "Apocalyptic AI: Religion and the Promise of Artificial Intelligence." *Oxford University Press, Journal of the American Academy of Religion,* 76, 1 (2008): 138–166.

———. "Robots and the Sacred in Science and Science Fiction: Theological Implications of Artificial Intelligence." *Zygon* 42, no. 4 (December 2007): 961–978. https://onlinelibrary.wiley.com/doi/epdf/10.1111/j.1467-9744.2007.00883.x?saml_referrer (23/01/2021).

———. *Virtually Sacred: Myth and Meaning in* World of Warcraft *and* Second Life. Oxford: Oxford University Press, 2014.

German Federal Ministry of Education and Research, the Federal Ministry for Economic Affairs and Energy, and the Federal Ministry of Labour and Social Affairs. "Artificial Intelligence Strategy." 2018, n9. https://www.ki-strategie-deutschland.de/home.html (06/12/2020).

Germanus. *Ecclesiastical History and Secret Theory* 1, *PG* 98, 384B.

Giannoulatos, Anastasios, Archbishop of Tirana, Durrës and All Albania. "Orthodox Views about the Development of Science and Technology." Welcome Address, November 15, 2020. https://orthodoxalbania.org/2020/el/2020/11/15/%ce%bf%cf%81%ce%b8%cf%8c%ce%b4%ce%bf%ce%be%ce%b5%cf%82-%ce%b1%cf%80%cf%8c%cf%88%ce%b5%ce%b9%cf%82-%cf%83%cf%87%ce%b5%cf%84%ce%b9%ce%ba%ce%ac-%ce%bc%ce%b5-%cf%84%ce%b7%ce%bd-%ce%b1%ce%bd%ce%ac%cf%80/ (17/01/2021).

Gkrekas, Aristarchos Archim. *The Church as a Place of Communion between God, Man and the World.* Thessaloniki: Ostracon Publishing p.c., 2014.

———. *Patristic Literature and Theology I.* Thessaloniki: Ostracon Publishing p.c., 2020.

Gregoire De Nysse. *La création de l'homme*, *SC* 6. J. Laplace. Paris-Lyon: 1943.

Gregorius, Nazianzenus, Theol. *In sanctum pascha (orat. 45)* (2022.052) Volume 36, page 633, line 51.

Gregorius, Nyssenus Theol. *De opificio hominis* (2017.079) (A.D.4), Page 185 line 32.

Gregory of Nazianzus. *Epitaph to Caesarion, Discourse* 7,23, *PG* 35,785.

———. *On the Epiphany*, Discourse LH, 11, *PG* 36, 321C–324B.
———. *Theological Oration V*, Discourse LA, *PG* 36, 169A–172A.
———. *Discourse*, 38,11, *PG* 36, 317.
———. *Discourse*, 41,9, *PG* 36, 441.
———. *Epistle to Cledonion* 101, *PG* 37,181.
Gregory of Nyssa. *The Complete Works*, I. Dogmatic treatises A', Series Greek Church Fathers, transl. Athenagoros Kokkinakis, Archbishop of Thyateira and Great Britain, ed. *GCF* Patristic Publications "Gregory Palamas." Thessaloniki: 1979; vol. 5, Thessaloniki: 1987.
———. *Catechetical Discourse*, 5, Srawley.
———. *Catechetical Discourse, Writings, GCF* vol. 1. Thessaloniki: 1979.
———. *On the Making of Man*, Writings, *GCF* 5. Thessaloniki: 1987, p. 208.
———. *Against Apollinarius*, 54, *PG* 45, 1253C–1256A.
———. *Homily on prayer*, 4, *PG* 44, 1165AB.
———. *On the Song of Songs*, 13 *PG* 44, 772A–1048C.
———. *Homily on the Romans*, 13,8, *PG* 60, 519.
———. *On the soul and resurrection*, *PG* 46, 60D–101B.
———. *Catechetical Discourse*, *PG* 45, 25D–93A.
———. *On the Making of Man*, *PG* 44, 133CD–240.
———. *On Perfection*, *PG* 46, 285C.
———. *On What It Is "in the Image" and "in the Likeness" of God*, *PG* 44,1336–1344. 1336AB–1340CD.
———. *On the Divine Purpose*, *PG* 46, 289C.
———. *On the Soul and the Resurrection*, *PG* 46, 12–232A.
———. *Let Us Make Man in Our Image and Likeness*, *PG* 44, 273AB–280D.
Gregory of Sinai. *Philokalia of the Holy Neptic (Fathers)*, vol. D'. Athens: Astir, 1991.
Gregory, Palamas. *Disputation to Akindynos*, vol. 3. P. Christou, Thessaloniki: 1970.
———. *Writings*, vol. 1, Thessaloniki 1962, vol. 2. Thessaloniki: 1970.
———. *Epistle to Akindynos* 3,16. P. Christou.
———. *Homily* 52, 59, 60. S. Oikonomou, Athens: 1861.
———. *In Favour of the Holy Resting* 1,3,26, 1,3,43. P. Christou.
———. *On the Divine Methexis* 19, ed. P. Christou, Gregory Palamas, *Writings*.
———. *Demonstrative Discources on the Procession of the Holy Spirit*, vol. 2.
———. *Chapters*, 61, *PG* 150, 1165A.
———. *Homily 3*, *PG* 151, 37A.
———. *Homily*, 16, *PG* 151, 189B.
———. *Homily XI, On the Precious and Life-Giving Cross*, 18, *PG* 151, 137B.
———. *Homily 19*, *PG* 151, 196A-261D.
———. *Homily 21*, *PG* 151, 277AB.
———. *Homily 34*, *PG* 151, 432D.
———. *Homily XXI*, *PG* 151, 277AB.
Gregory the Theologian. *On Holy Easter*, *PG* 36, 632AB.
———. *Discourse G, Epitaph to Brother Caesarion*, XXIII, *PG* 35,785C.

Grossman, Lev. "2045: The Year Man Becomes Immortal." *Time,* February 10, 2011. http://content.time.com/time/magazine/article/0,9171,2048299-4,00.html (03/01/2021).

Gutkind, Lee. *Almost Human: Making Robots Think*. New York: W.W. Norton, 2006.

G20 Trade Ministers and Digital Economy Ministers. "G20 Ministerial Statement on Trade and Digital Economy." 2019, Principle 1.2. https://www.mofa.go.jp/files/ 000486596.pdf (03/03/2021).

Hambling, David. "Drones May Have Attacked Humans Fully Autonomously for the First Time." *newscientist*, May 27, 2021. https://www.newscientist.com/article/2278852-drones-may-have-attacked-humans-fully-autonomously-for-the-first-time/ (17/07/2021).

Harari, Yuval. *Homo Deus: A Brief History of Tomorrow*. New York: Harper, 2017.

Harris, Mark. "Inside the First Church of Artificial Intelligence." *Wired Magazine,* 2017. https://www.wired.com/story/anthony-levandowski-artificial-intelligencereligion/ (10/12/2020).

Harrison, Brent, and Mark Riedl. "Towards Learning from Stories: An Approach for Interactive Machine Learning." Paper presented at the AAAI'16 Workshop on Symbiotic Cognitive Systems, 2016, 746–750. https://www.cc.gatech.edu/:riedl/pubs/aaai-symbiotic16.pdf (14/01/20210).

Harrison, Verna. "Receptacle Imagery in St. Gregory of Nyssa's Anthropology." *SP 22* (Leuven 1989): 23–27.

Harrys, Tom. "How Robots Work." *howstuffworks.com.* https://science.howstuffworks.com/ robot4.htm (12/12/ 2020.

Hendricks, Beth. *Intelligent Agents: Definition, Types & Examples.* https://study.com/academy/lesson/intelligent-agents-definition-types-examples.html (11/12/2020.

Hoffman, Chris. *15 Things You Can Do With Cortana on Windows 10*, 2015. https://www.howtogeek.com/225458/15- things-you-can-do-with-cortana-on-windows-10/ (10/12/ 2020.

Hunt, Hannah. "Penthos and Repentance in St. Symeon the New Theologian." *Studia Patristica* 35 (2001): 114–120.

IEEE Global Initiative on Ethics of Autonomous and Intelligent Systems. "Ethically Aligned Design: A Vision for Prioritizing Human Well-Being with Autonomous and Intelligent Systems." (2019) First Edition. https://ethicsinaction.ieee.org,/ (03/12/2020).

Iefimerida. *A New Religion Is Born: The God of Artificial Intelligence.* 29/11/2017. iefimerida.gr-https://www.iefi merida.gr/news/379554/mia-nea-thriskeia-gennithike-o-theos-tis-tehnitis-noimosynis-eikones (12/12/2020).

Information Technology Industry Council. "AI Policy Principles." 2017, n 9, p. 5 (05/12/2020).

Irenaeus of Lyons. *Control*, 5,6,1, *PG* 7, 1137A.

———. *Against Heresies*, 5,6,1, *PG* 7, 1158 A.C.

Istvan, Zoltan. "When Superintelligent AI Arrives, Will Religions Try to Convert It?" *Gizmodo,* 2/04/15. https://gizmodo.com/when-superintelligent-ai-arrives-will-religions-try-t-1682837922/ (03/04/2021).

Japanese Cabinet Office, Council for Science, Technology and Innovation. "Social Principles of Human-Centric Artificial Intelligence." n 20, 2019. https://www8.cao.go.jp/cstp/english/human-centricai.pdf (06/02/2021).

John Chrysostom. *Goths Reading, and a Gothic Elder,* spoken in the Church of Paul, 8, *PG* 63, 508–510.

———. *On the Ascension*, 16, *PG* 52,789.

———. *On the Ambiguities of the Old Testament*, 2,5, *PG* 56, 182

———. *Homily on First Thessalonians* 10, 3, *PG* 62, 458.

———. *Homily on First Corinthians*, 17, 3, *PG* 61, 143.

———. *On First Timothy, Homily* 7, 2, *PG* 62, 536.

———. *On Genesis, Homily*, *PG* 53,71-293, *PG* 54, 590.

———. *Homily On Hebrews*, 6,2, *PG* 63,56.

———. *Homily on Ephesians* 3,1, *PG* 62-118, 27–1221.

———. *Homily on Romans*, 11, 2, *PG* 60, 174–486.

———. *Homily on Romans* 8, 2, *PG* 60, 510.

———. *Psalm 48*, 7, *PG* 55, 233.

———. *On the Gospel of John PG* 59, 159.

———. *Commentary on the Psalms, PG* 55, 469.

———. *Homily on the Holy Pentecost*, *PG* 50, 456.

———. *Psalm 144* 1, *PG* 55, 465.

———. *On "My Grace Is Sufficient for You."* 2, *PG* 59, 509.

———. *On St. John*, 2,5, *PG* 59, 36.

———. *Discourse on Blessed Babylas* 2, *PG* 50, 536.

———. *On the Earthquake and the Rich Man and Lazaru*s 7, *PG* 48, 1037–1038.

———. *On Christ's Two Wills*, 30, *PG* 95, 168.

———. *On the Priesthood*, 2,3, *PG* 48, 634.

———. *To Stageirion*, 1, 2, *PG* 47, 428.

———. *Homily. After Eutropius Having Been Found Outside the Church Had Been Taken Captive, and on Paradise and the Scriptures*, *PG* 52, 14–18.

———. *To the Scandalised* 1, *PG* 94, 984.

———. *Before the Exile*, *PG* 52, 35.

———. *Homily on Pentecost* 1, 2, *PG* 50, 456.

John Damascene. *Bible of Wuestions*, *PG* 90, 1109D–112A.

———. *Dialectic*, *PG* 94, 609A–616B.

———. *An Exact Exposition of the Orthodox Faith*, Introduction, comments, and translation by Matsoukas Nikolaos.. Kyriakidis, Thessaloniki: 2020.

———. *An Exact Exposition of the Orthodox Faith*, *PG* 94, 320–1153.

———. *Against Manichaeus*, *PG* 94, 1548A–1548A.

———. *On Christ's Two Wills*, *PG* 95, 148–149A.

———. *On Holy Images*, 1, *PG* 94, 1300AB.

———. *Against Those Who Decry the Holy Images C', PG* 94, 1340D–1341.

John Sinaites. *The Ladder*, 29, *PG* 88, 1129A–1149A.

Jurafsky, Daniel, and James Martin. *Speech and Language Processing: An Introduction to Natural Language Processing, Computational Linguistics and*

*Speech Recognition*, 2020. https://web.stanford.edu/:jurafsky/slp3/ed3book.pdf (17/12/2020).

Kallistos, Patriarch. "Chapters on Prayer." *Philokalia of the Holy Neptic (Fathers)*, vol. D'. Athens: Astir, 1991.

Kallistos, Katafigiotis. "The Surviving Documents on the Divine Union and Theoretical Life." *Philokalia of the Holy Neptic (Fathers)*, vol. E'. Athens: Astir, 1992.

Kalogeros, Ioannis. *On the Character of the Orthodox Catholic Church According to the Fundamental Soteriological Principles of the New Testament*, offprint of the honorary volume on the Jubilee of His Eminence Metropolitan of Philippi, Neapolis and Thassos Chrysostom. Athens: 1961.

Karavidopoulos, Ioannis. *God's Icon and "in the Image" of God by the Apostle Paul. The Christological Foundations of Pauline Anthropology.* Thessaloniki: 1964.

Karmiris, Ioannis. *Orthodox Ecclesiology, Dogmatics Part E'.* Athens: 1973.

———. *Christian Ethics*, vol. A'. Athens: 1967.

———. *The Ecclesiology of the Three Hierarchs.* Athens: 1962.

———. *Synopsis of the Doctrine of the Orthodox Catholic Church.* Athens: 1960.

Kelly, Kevin. "What's the Relationship of Technology to God?" *Religion Dispatches*, June 16, 2016. https://religiondispatches.org/ kevin-kelly-interview/(03/03/2021).

Kok, Joost Nico. *Artificial Intelligence.* Oxford, UK: Eolss/UNESCO, 2009.

Krasadakis, Georgios. "Artificial Intelligence: What It Is and How It Is Drastically Changing Our World." 3rd Naftemporiki Technology Conference, *Artificial Intelligence: A Blessing or a Threat*, Royal Olympic Hotel, January 28, 2018. Athens: https://www. naftemporiki.gr/story/1309187/texniti-noimosuni-ti-einai-kai-pos-allazei-dramatika-ton-kosmo-mas (10/01/2021).

Kurzweil, Ray. *The Age of Spiritual Machines.* New York: Viking, 1999.

———. *The Singularity Is Near.* New York: Viking, 2005.

Laird, Harris, Gleason Archer, and Bruce K. Waltke. *Theological Wordbook of the Old Testament 2*. Chicago: Moody Press, 1980.

Legrand, Philippe-Ernest. *St. J. Chrysostome, Les Moralistes Chrétiens*. J. Gabalda. Paris: 1924.

Lekkas, Georgios, Archpriest, Counsellor at the Representation Office of the Church of Greece to the E.U. "Artificial Intelligence and an Applied Ethics. But What Kind of Ethics?" Brussels, 21.1.2019. https://ec.europa.eu/futurium/en/european-ai-alliance/artificial-intelligence-and-applied-ethics-what-kind-ethics (19/12/2020).

Le, Michelle. *Inside the First Church of Artificial Intelligence*, Wired, 11.15.2017 https://www.wired.com/story/anthony-levandowski-artificial-intelligence-religion/ (16/12/2020).

Lenat, Douglas. "Cyc: A Large-Scale Investment in Knowledge Infrastructure." *Communications of the ACM* 38 (1995): 33–38.

Leontius, Byzantius. *Against Nestorius and Eutyches*, 2, *PG* 86, 1349A.

Levchuk, Kate. "AI vs. God: Who Stays and Who Leaves?" *Forbes*, August 5, 2018 https://www.forbes.com/sites/cognitiveworld/ 2018/08/05/ai-vs-god-who-stays-and-who-leaves/?sh=4ab0809f2713 9 (02/03/2021).

Lossky, Vladimir. *The Mystical Theology of The Eastern Church*. Crestwood, NY: St Vladimir's Seminary Press, SPS, 1976.

———. *The Vision of God*, trans. M. Kalamaras.. Rigopoulos, Thessaloniki: 1973.

———. *The Vision of God*, ed. Holy Metropolis of Nicopolis. Preveza: 2004.

———. *The Mystical Theology of the Eastern Church*. Thessaloniki: 1986.

Loudovikos, Nikolaos Father. *A Eucharistic Ontology*. Domos, 1992.

Luckin, Rose. "Towards Artificial Intelligence-Based Assessment Systems." *Nature Human Behaviour* volume 1, Art.n. 0028, 2017. 1–3. https://www.nature.com/articles/s41562-016-0028?proof=tNature-doi:10.1038/s41562–016–0028 (18/12/2020).

Mantzaridis, George. *Christian Ethics I, Introduction-General Principles, Modern Challenges*. Pournara, Thessaloniki: 2004.

———. *Christian Ethics II, Man and God, Man and Fellow Man, Existential and Bioethic Propositions and Perspectives*. Pournara, Thessaloniki: 2004.

Mark the Hermit. *On Repentance*, 1, *PG* 65, 965BC.

———. *Of the Spiritual Law*, 196, *PG* 65, 929A.

———. *Concerning Those Who Think to be Justified through Works*, 33, *PG* 65, 936B.

Marsselos, Alexandros. "Basil the Great as a Physician." *Vassilias, Festive Volume on the Completion of 1600 Years since the Death of Basil the Great*. Holy Metropolis of Thessaloniki. Thessaloniki: 1979.

Matsoukas, Nikolaos. *Dogmatic and Symbolic, Theology A'. Introduction to Theological Epistemology*. Pournara, Thessaloniki: 1985.

———. *Dogmatic and Symbolic Theology B'*. Pournara, Thessaloniki: 2001.

———. *Dogmatic and Symbolic Theology B', Exposition of the Orthodox Faith*. Kyriakidi, Thessaloniki: 2016.

———. *Dogmatic and Symbolic Theology B'. Exposition of the Orthodox Faith in Contrast with Western Christianity, Philosophical and Theological Library 3*. Pournaras, Thessaloniki: 2001.

———. *Dogmatic and Symbolic Theology, C'. Recapitulation and Agathotopia*. Exposition of the ecumenical character of Christian teaching. Pournara, Thessaloniki: 1997.

———. *Dogmatic and Symbolic Theology D', Satan*. Pournara, Thessaloniki: 1999.

———. *Theology, Ktisiology, Ecclesiology according to Athanasius the Great. Points of Patristic and Ecumenical Theology*. Pournara, Thessaloniki: 2001.

———. *Cosmos, Man, Society according to Maximus the Confessor*. Grigoris, Athens: 1980.

———. *The Problem of Evil. Essay on Patristic Theology*. Pournara, Thessaloniki: 1986.

Maximus the Confessor. *Mystagogy*. Apostoliki Diakonia, Athens: 1973.

———. *Letters XVI, To Him on Slander and Redemption by Divine Assistance*, *PG* 91, 557D–560A.

———. *Letter 2*, *PG* 91, 396CD.

———. *Disputations with Pyrrhus*, *PG* 91, 309C.

———. *Chapters on Theology*, *PG* 90, 1136.

———. *On Various Questions*, *PG* 91, 1305B–1048B.

———. *To Thalassios on Difficulties in Sacred Scripture*, *PG* 90, 44D–436.
———. *Various Chapters* 5,93, *PG* 90, 1193D–1388CD.
———. *Various Theological and Economic Chapters*, *PG* 90,1344.
———. *Chapters on Love*, 4,55, *PG* 90, 993A–1060C.
———. *Various Chapters on Theology and Economy*, 2,71, *PG* 90, 1101AB–1156D.
———. *Mystagogy*, 24, *PG* 91, 701D–704A.

McCorduck, Pamela. *Machines Who Think: A Personal Inquiry into the History and Prospects of Artificial Intelligence*. 2nd ed. Natick, MA: A K Peters/CRC Press, 2004.

McGrath, James. *Did Jesus Claim to Be God?* September 2016. https://www.patheos.com/blogs/religionprof/2016/09/siri-pray-for-me.html (10/01/2021).

McHargue, Mike. *Finding God in the Waves: How I Lost My Faith and Found It Again through Science.* Colorado: Convergent Books, 2016.

Merritt, Jonathan. "Is AI a Threat to Christianity? Are You There, God? It's I, Robot." Religion Dispatches, February 3, 2017. https://religiondispatches.org/kevin-kelly-interview/ (20/02/2021).

Meyer, Louis. *St. J. Chrysostome, Maître de perfection chrétienne.* Paris: Gabriel Beauchesne et Ses Fils, 1933.

Microsoft. "AI Principles." 2018. https://www.microsoft.com/en-us/ai/our-approach-toai (09/02/2021).

Miller, Erik. "Introduction to Computer Vision." University of Massachusetts Amherst, 2011. https://people.cs.umass.edu/:elm/Teaching/Docs/IntroCV_1_19_11.pdf (19/12/2020).

Mills, Frederick, and Robert Stufflebeam. *Introduction to Intelligent Agents.* 2005. https://miatedjosaputro.com/uploads/sites/2/2020/05/Introduction-to-Intelligent-Agents.pdf (12/12/20200.

Minsky, Marvin. *Society of Mind.* New York: Simon and Schuster, N1985.

Mission assigned by the French Prime Minister. "For a Meaningful Artificial Intelligence: Toward a French and European Strategy." 2018. https://www.aiforhumanity.fr/pdfs/MissionVillani_Report_ENG-VF.pdf (10/12/2020).

Modern, John. *Neuromatic; or, a Particular History of Religion and the Brain.* December 2020. http://www.johnmodern.com/2020/12/15/neuromatic-or-a-particular-history-of-religion-and-the-brain/ (05/03/2021).

Money, Control. *Elon Musk Slams Proposal to Develop an Artificial Intelligence God.* October 25, 2017. https://www.moneycontrol.com/news/technology/elon-musk-slams-proposal-to-develop-an-artificial-intelligence-god-2419585 (12/12/2020).

Moor, James. *The Nature, Importance, and Difficulty of Machine Ethics*. IEEE Intelligent Systems, 2006. Doi: 10.1109/MIS. 2006.80 (11/12/2020).

Moravec, Hans. *Robot: Mere Machine to Transcendent Mind.* New York: Oxford University Press, 1999.

———. "Letter from Moravec to Penrose." In *Thinking Robots, an Aware Internet, and Cyberpunk Librarians: The 1992 LITA President's Program*, edited by R. Bruce Miller and Milton T. Wolf, 51–58. Chicago: Library and Information Technology Association, 1992a.

———. *Mind Children: The Future of Robot and Human Intelligence*. Cambridge, MA: Harvard University Press, 1988.

———. *Robot: Mere Machine to Transcendent Mind*. New York: Oxford University Press, 1999.

Moutsoulas, Elias, *The Arian Dispute and Athanasius the Great*, Part B′, *Athanasius the Great*, vol. A’. Athens: 1978.

———. *The Incarnation of the Word and the Theosis of Man according to the Teaching of Gregory of Nyssa*. Athens: 1965.

Mumford, Lewis. *Technics and Civilization*. Chicago University of Chicago Press, 1934.

Musaddique, Shafi. “How Artificial Intelligence Is Shaping Religion in the 21st Century.” *Cnbc,* May 11, 2018. https://www. cnbc.com/2018/05/11/how-artificial-intelligence-is-shaping-religion-in-the-21st-century.html (14/12/2020).

NBC News. *Pepper The Robotic Buddhist Priest Debuts In Japan.* https://www.youtube.com/watch?v=FVobokmWqe8, https://www.religiousstudiesproject.com/podcast/artificial-intelligence-and-religion/ (27/12/2020).

Nellas, Panayiotis. *Devinely Inspired Animal, Perspectives for an Orthodox Understanding of Man*. Athens: Epopeia, 1979.

Nelson, Victoria. *The Secret Life of Puppets*. Cambridge, MA: Harvard University Press, 2001.

Nicholas, Kabasilas. *Commentary on the Divine Liturgy*, 22, *PG* 150,146.

———. *On the Life in Christ*, *PG* 150, 493–680AB.

Nicodemos the Hagiorite. *Eortodromion (or Commentary on the Canons of the Dominical and Marian Feasts)*, vol. B. Orthodoxos Kypseli. Thessaloniki: 1995.

———. *Garden of Graces (or an Elegant Interpretation of the Nine Odes of the Stichologia)*. Rigopoulos, Athens: 1992.

———. *New Klimax (or Commentary on the Seventy-FIve Anavathmoi of the Octoechos)*. Schinas, Volos, 1956.

———. *The Rudder (Pedalion) of the Metaphorical Ship of the One, Holy, Catholic and Apostolic Church of the Orthodox Christians*, 8th reprint of the 3rd edition of 1846.. Rigopoulos, Athens: 1976.

———. *A Handbook of Spiritual Counsel*. Panagopoulos, Athens: 2001.

———. *Ethics of Christians*. Rigopoulos, Thessaloniki: 1998.

Nikolaidis, Apostolos. *Social Ethics*. Grigoris, Athens: 2009.

Noble David. *The Religion of Technology: The Divinity of Man and the Spirit of Invention*. New York: Knopf Doubleday Publishing Group, 2013.

Official Documents of the Holy and Great Council of the Orthodox Church. *Encyclical V 11. The Church in the Face of Globalization, Extreme Violence and Migration § 16*, Crete, June 2016 https://www.holycouncil.org/-/encyclical-holy-council (19-11-2020).

———. *Encyclical I.* F §1, Crete. June 2016. https://www.holycouncil.org/-/encyclical-holy-council (19-12-2020).

———. *Message,* §7, Crete. June 2016. https://www.holycouncil.org/-/message (11-11-2020).

Organisation for Economic Co-operation and Development. "Recommendation of the Council on Artificial Intelligence." (n 54) https://legalinstruments.oecd.org/en/instruments/OECD-LEGAL-0449 (03/03/2021).

Otto, Rudolph. *The Idea of the Holy.* Translated by John W. Harvey. Oxford: Oxford University Press, 1923, 2nd ed., 1950; Das Heilige, 1917.

———. *The Idea of the Holy.* Translated by J. W. Harvey. London: Oxford, [1917] 1923.

Panagiotopoulos, Themistocles, and Georgios Anastasakis. "Logic Programming." University of Piraeus, 2012.

Panagopoulos, Ioannis. "Ontology or Theology of the Person. The Contribution of Patristic Trinitarianism to the Understanding of the Humans." *Synaxis* 14 (1985): 35–47.

Pannenberg, Wolfhart, *Was ist der Mensch? Die Anthropologie der Gegenwart im Lichte der Theologie*. Göttingen: Vandenhoeck & Ruprecht, 1976.

Papadopoulou-Tsanana, Olympia. *The Anthropology of Basil the Great*, ed. Patriarchal Foundation for Patristic Studies. Thessaloniki: 1970.

Papathanasiou, Konstantinos father. *Salvation in Johannine Theology*. Pournara, Thessaloniki: 2010.

Peter of Damascus. "On the Seven Bodily Acts." *Philokalia of the Holy Neptic (Fathers)*, vol. C'. Athens: Astir, 1991.

Plaisant, Catherine, Maxine Cohen, Steven Jacobs, Niklas Elmqvist, and Nicholas Diakopoulos. *Designing the User Interface: Strategies for Effective Human-Computer Interaction*, Sixth Edition. Boston: Pearson, 2016.

Plato. *Gorgias* 456B.

———. *Politeia*, VI,492B.

Plexidas, Ioannis. *The Anthropology of Evil, the Problem of Theodicy and Its Anthropological Presuppositions in the Thought of Gregory of Nyssa.* Thessaloniki: 2010. https: //gunet2.cs.unipi.gr/modules/document/file.php/TME137/%CE%A3 %CE%B7%CE%BC%CE%B5%CE%B9%CF%8E%CF%83%CE%B5%CE%B 9%CF%82/notes-lp-msc-inf-2012.pdf (15/12/2020).

Polybius. *Histories*, 5,74.

Prisco, Giulio. "Reminder: Join the Turing Church Discord Server!." Turingchurch 2016. https://turingchurch.net/about-turing-church-ac6ebf2e97b6/ (10/12/2020).

Protothema. *Google's Engineer Wants to Develop an Artificial Intelligence Deity*, 29/09/2017. https://www.protothema.gr/technology/article/ 717804/mihan-ikos-tis-google-thelei-na-anaptuxei-mia-theotita-tehnitis-noimosunis/ (15/12/2020).

Rallis, Georgios. *Constitution of the Divine and Sacred Canons of the Holy and Praised Apostles and of the Sacred Ecumenical and Local Councils and of the Local Holy Fathers*, from the printing press of Hatofilakos. Potlis Michail, Athens: 1852–1859. C', 561 https://anemi.lib.uoc. gr/metadata/9/3/c/metadata-01-0000374.tkl (10/02/2021).

Raso, Filippo, Hannah Hilligoss, Vivek Krishnamurthy, Christopher Bavitz, and Levin Kim. "Artificial Intelligence & Human Rights: Opportunities and Risks." Berkman Klein Center Research Publication, September 25, 2018, 58. https://cyber.harvard.edu/sites/default/files/2018-09/2018-09_AIHumanRightsSmall.pdf (03/12/2020).

Remington, Tonar. *The Church of Artificial Intelligence: A Religion in Need of a Responsible Theology.* August 2017. https://medium.com/@AItheist/the-church-of-artificial-intelligence-a-religion-in-search-of-a-theology-fa7e73f46900 (12/12/2020).

Rendsburg, Melissa. *The Impact of Artificial Intelligence on Religion: Reconciling a New Relationship with God.* July 2019. https://www.academia.edu/40235300/The_Impact_of_Artificial_ Intelligence_ on_Religion_Reconciling_a_New_Relationship_with_ God (17/01/2021).

Riedl, Mark. "Computational Narrative Intelligence: A Human-Centered Goal for Artificial Intelligence." Georgia Institute of Technology. Atlanta: 2016. https://arxiv.org/pdf/1602.06484.pdf (14/01/ 2021).

———. "Human-Centered Artificial Intelligence and Machine Learning." Wiley, School of Interactive Computing, Georgia, Institute of Technology. Atlanta: 2018. DOI: 10.1002/hbe2.117 (18/12/2020).

Riedl, Mark, and Brent Harrison. "Using Stories to Teach Human Values to Artificial Agents. In Proceedings of the 2nd International Workshop on AI, Ethics and Society." Georgia Institute of Technology. Atlanta: 2016. https://www.cc.gatech.edu/:riedl/pubs/aaai-ethics 16.pdf (17/01/20210).

Robert, Lionel. "The Growing Problem of Humanizing Robots." *International Robotics & Automation Journal* 3, no. 1 (2017): 247–248.

Rory, Cellan-Jones. "Stephen Hawking Warns Artificial Intelligence Could End Mankind." BBC, December 2, 2014 https://www.bbc.com/news/technology-30290540 (20/02/2021).

Rothblatt, Martine. *The Incredible Polymath of Polymaths.* https://www.youtube.com/watch?v=S1rExMw-13A/ (03/03/2021).

———. *Virtually Human: The Promise—and the Peril—of Digital Immortality*. New York: St. Martin's Press, 2014.

Roy, Jessica. "The Rapture of the Nerds." *Time,* April 17, 2014. https://time.com/66536/terasem-trascendence-religion-technology/ (20/03/2021).

Rusanos, Pachomius. "To Dionysius." Karmiris Ioannis, *Pachomius Rusanos and the unpublished dogmatic and other works of him*, Texte und Forschungen zur byzantinischneugriechischen Philologie, 14+. Athens: 1935.

Russell, Bjork. *Artificial Intelligence and the Soul.* 2008. https://www.asa3.org/ASA /PSCF/2008/PSCF6-08Bjork.pdf (02/02/20210.

Russell, Stuart. *Compatible: Artificial Intelligence and the Problem of Control.* New York: Penguin Random House, 2019.

———. "Take a Stand on AI Weapons." *Nature* 521 (May 2015): 415–418. https:// www.nature.com/articles/521415a. pdf?origin=ppub (15/11/2022).

Russell, Stuart, and Peter Norvig. "Artificial Intelligence: A Modern Approach." Translation by Takis Alvas, Dimitris Kartsaklis, and Fotis Skoularikis. Athens: Klidarithmos, 2005.

Sadeghi, Fereshteh, Divvala Santosh, and Farhadi Ali. "VisKE: Visual Knowledge Extraction and Question Answering by Visual Verification of Relation Phrases." Paper presented at IEEE Conference on Computer Vision and Pattern Recognition (CVPR) (2015) 1456–1464.

Sakharov, Archim Sophrony. *St. Silouan the Athonite*. Monastery of Saint John the Baptist. Essex, England: 1999.

Savvatos, Chrysostomos, metropolitan bishop of Messenia. "Man and the Evolution of Modern Technology. A Challenge for Orthodox Theology." Holy Metropolis of Messinia. Kalamata: 2019.

———. "Man and the Evolution of Modern Technology. A Challenge for Orthodox Theology, the Technology Phenomenon." I, *Theology*, 90, 3 (2019): 19–41.

Schneier, Bruce. "Resources on Existential Risk—for: Catastrophic Risk: Technologies and Policies." Berkman Center for Internet and Society, Harvard University, 2015 https://futureoflife.org/data/documents/Existential%20Risk%20Resources%20(2015-08-24).pdf?x70892. (18/12/2020).

Schoepflin, Rennie. "Politics and Apocalypticism." Apocalypticism in the Modern World and the Contemporary Age. New York: Continuum, 2000.

Sekiguchi, Kaira, and Hori Koichi. "Organic and Dynamic Tool for Use with Knowledge Base of AI Ethics for Promoting Engineers' Practice of Ethical AI Design." *Springer, AI & Society, Journal of Knowledge, Culture and Communication*, 35 (2020): 51–71. https://link.springer.com/article/10.1007/s00146-018-0867-z (10/11/2020).

Sheridan, Thomas. "Function Allocation: Algorithm, Alchemy or Apostasy?." *International Journal of Human-Computer Studies* 52, no. 2 (2000): 203–216.

Sherwood, Harriet. "Robot Priest Unveiled in Germany to Mark 500 Years since Reformation." *The Guardian,* May 2017. https://www. theguardian.com /technology/2017/may/30/robot-priest-blessu-2-germany-reformation-exhibition (03/01/2021).

Shneiderman, Ben. "Bridging the Gap between Ethics and Practice: Guidelines for Reliable, Safe, and Trustworthy Human-Centered AI Systems." *ACM Transactions on Interactive Intelligent Systems,* October 2020c https://dl.acm.org/doi/10.1145/3419764 (03/01/2021).

———. *Designing the User Interface: Strategies for Effective Human-Computer Interaction*, ed. Addison-Wesley, Boston 1987.

———. "Design Lessons from AI's Two Grand Goals: Human Emulation and Useful Applications." *IEEE Transactions on Technology and Society* 1, no. 2 (June 2020b): 73–82.

———. "Human-Centered Artificial Intelligence: Reliable, Safe & Trustworthy." *International Journal of Human-Computer Interaction* 36, no. 6 (March 2020a): 495–504.

———. "Human-Centered Artificial Intelligence: Three Fresh Ideas." *AIS Transactions on Human-Computer Interaction*, V.12 I.3 A.1, 9–30-2020, 109–123.

———. "Opinion: The Dangers of Faulty, Biased, or Malicious Algorithms Requires Independent Oversight." *Proceedings of the National Academy of Sciences* 113, no. 48 (2016): 13538–13540.

Skaltsas, Georgios. "Man as a Mirror of the Eschaton according to Saint Gregory of Nyssa." *Synaxis* 59 (1996): 45–49.

Skouteris, Konstantinos. *Man According to the Word of the Divine Economy.* http://www.apostoliki-diakonia.gr/bible/bible.asp?contents=themata/contents.asp&main=diafora&file=2.2.htm (08/01/2021).

Stamoulis, Chrysostomos. *About Light. Personal or Physical Energies? Contribution to the Modern Problematic Holy Trinity in the Orthodox Space.* To Palimpsiston, Thessaloniki: 1999.

———. *Cyril of Alexandria, On the Incarnation of the Only-Begotten.* To Palimpsiston, Thessaloniki: 1998.

Stăniloae, Dimitru, Father. *God, the World and Man.* Athens: Armos, 1990.

———. *The Orthodox Christian Teaching on God.* Athens: Armos, 2011.

———. *The Apostolic Canons*, 34, Literature of the Greek Church Fathers and Theological Authors (LGCFTA), vol. 2. Apostoliki Diakonia, Athens: 1955.

Stark, Rodney, and William Sims Bainbridge. *A Theory of Religion.* Berne, New York: Peter Lang, 1987.

Stone, Allucquere Rosanne. *Will the Real Body Please Stand Up?: Boundary Stories about Virtual Cultures.* MIT Press, Cambridge 1991.

Strauss, Mark. "Would Artificially Intelligent Androids Be Religious Fundamentalists?" *Gizmodo,* 9/03/14 https://io9.gizmodo.com/would-artificially-intelligent-androids-be-religious-fu-1630137541/(05/02/2021).

Stuart, Russell, Daniel Dewey, and Max Tegmark. "Research Priorities for Robust and Beneficial Artificial Intelligence." *Future of Life Institute - AI Magazine,* 2015, 105–114 https://futureoflife.org/data/documents/research_priorities.pdf?x90991 (07/12/2020).

Summers, Jason. "Can Singularity and Christianity Coexist?" *Think Christian,* April 12 2011 https://thinkchristian.net/can-singularity-and-christianity-coexist(03/02/2021).

Suzuki, Kenji. *Artificial Neural Networks: Architectures and Applicatio*ns, ed. Intechopen, Croatia 2013.

Symeon of Thessalonica. *On the Holy Temple*, CXXXVI, *PG* 155, 345C-345D.

Symeon the New Theologian. *Ethical Discourses*, 10,11-35, ed. J. Darrouzes. "Sources Chrétiennes." vol. 129. Paris: 1967.

———. *The Secret Wish, SC* 156.

———. *The Ethical Discourses* vol. 1, *SC* 122.

———. *The Ethical Discourses* vol. 2, *SC* 122.

———. *The Ethical Discourses* vol. 3, *SC* 122.

———. *The Ethical Discourses* vol. 4, *SC* 129.

———. *The Ethical Discourses* vol. 5, *SC* 129.

———. *The Ethical Discourses* vol. 6, *SC* 129.

———. *The Ethical Discourses* vol. 8, *SC* 129.

———. *The Ethical Discourses* vol. 10, *SC* 129.

———. *The Ethical Discourses* vol. 13, *SC* 129.

———. *The Discourses* vol. 1, *SC 122.*

———. *The Discourses* vol. 2, *SC* 122.

———. *Catechesis* 5, 395–397, *SC* 96.
———. *Catechesis* 14, 122–123, *SC* 104.
———. *Catechesis 15*, 68–99, *SC* 104.
———. *Catechesis 16*, 91–92, *SC* 104.
———. *Catechesis* 20, 14–20, *SC* 104.
———. *Chapters* 3,3, *SC* 51.
———. *Catechisms*, 33, 10–16, *SC* 113.
———. *Hymn* 4, 58–64, *SC* 156.
———. *Hymn 23*, 448–455, *SC* 196.
———. *Hymn* 26, 42–50, *SC* 174.
———. *Hymn* 30, 463–466, *SC* 174.
———. *Hymn* 42, 188–192, *SC* 196.
———. *Hymn* 44, 30–54, *SC* 196.
———. *Hymn* 46, 27–37, *SC* 196.
———. *Hymn* 50, 203–212, *SC* 196.
———. *Hymn* 52, 1–6, *SC* 196.
———. *Hymn 53*, 112–132, *SC* 196.
———. *Hymn 55*, 142–147, *SC* 196.

Tencent Institute. "Six Principles of AI." n 58, Principle 1. http://www.kejilie.com/iyiou/article/ZRZFn2.html (06/02/2021).

Tenner, Edward. "Will Death Become Obsolete?" *The Atlantic,* February 12, 2011. https://www.theatlantic.com/technology/archive/2011/02/will-death-become-obsolete/71166/ (08/02/2021).

Terasem Movement Foundation. *About the Terasem Movement Foundation Inc.* https://te rasemmovementfoundation.com/mission (20/01/2021).

Tertullian. *De baptism 6*, Corpus Scriptorum Ecclesiasticorum Latinorum (CSEL), editume consilio et empensis Academiae Litterarum Caesareae Vindobonensis 1866, CSEL 20, 206.

Theoleptos of Philadelphia. *Philokalia of the Holy Neptic (Fathers)*, vol. D'. Athens: Astir, 1991.

Theodorou Andreas. *The Perceptions of Saint Gregory of Nazianzus Regarding the Human Body*, ΕΕΘΣΑ ΚΑ'. Athens: 1975.

———. *The Teaching of the Greek Fathers of the Church up to John of Damascene on the Theosis of Man.* Athens: 1956.

Thucydides. *History of the Peloponnesian War*, 1,31,87.

Toma, Stefan-Lucian. *The Patristic Tradition in the Work of Father Dimitru Stăniloae and the Modern World.* Pournara, Thessaloniki: 2007.

Trieu, H. Trinh, and Quoc V. Le. "A Simple Method for Commonsense Reasoning." *Cornel University* (2018) CoRR; abs/1806.02847. https://arxiv.org/abs/1806.02847 (14/01/2021).

Tselengidis, Dimitrios. *The Soteriology of Luther*. Pournara, Thessaloniki: 1991.

———. *Philokalia of the Holy Neptic (Fathers)*, vol. A.' Holy Monastery of the Assumption of Theotokos. Boura Arcadia: 2010.

———. *Grace and Freedom in the Patristic Tradition of the Eleventh Century*. Pournara, Thessaloniki: 1987.

Tsigos, Vassilios. *Dogmatics of the Orthodox Church*. Thessaloniki: Ostracon Publishing, 2017.

UK House of Lords, Select Committee on Artificial Intelligence. "AI in the UK: Ready, Willing and Able?" *Report of Session* 2017–19, 2018. https://publications.parliament.uk/pa/ld201719/ldselect/ldai/100/100.pdf (10/12/2020).

UNI Global Union. "Top 10 Principles for Ethical Artificial Intelligence." 2017, n 66, Principle 3. http://www.thefutureworldofwork.org/media/35420/uni_ethical_ai.pdf (03/02/2021).

United Nations, Secretary-General António Guterres. "UN Artificial Intelligence Summit Aims to Tackle Poverty, Humanity's 'Grand Challenges.'" 2017. https://news.un.org/en/story/2017/06/ (02/01/ 2021).

United Nations. "Universal Declaration of Human Rights." December 10, 1948. https://www.un.org/en/about-us/universal-declaration-of-human-rights (03/12/2020).

University of Montreal. "2018 Report Montréal Declaration for a Responsible Development of Artificial Intelligence." 2018. https://monoskop.org/images/b/b2/Report_Montreal_Declaration_for_a_Responsible_Development_of_Artificial_Intelligence_2018.pdf (06/02/2021).

US Public Policy Council (USACM), Association for Computing Machinery. "Statement on Algorithmic Transparency and Accountability." https://www.acm.org/binaries/content/assets/public-policy/2017_usacm_statement_algorithms.pdf (15/11/2020).

Vedantam, Ramakrishna, Xiaoyu Lin, Tanmay Batra, C. L. Zitnick, and Devi Parikh. "Learning Common Sense through Visual Abstraction." Paper presented at IEEE International Conference on Computer Vision (ICCV) (2015), 2542–2550.

Vlahavas, Ioannis. "Artificial Intelligence, Knowledge Systems." 2013, opencourses.auth.gr https://opencourses.auth.gr/modules/units/?course=OCRS118&id=1397 (12/12/20200.

Vlahavas, Ioannis, Petros Kefalas, Nikolaos Vasileiadis, Fotis Kokkoras, and Ilias Sakellariou. *Artificial Intelligence*. Thessaloniki: University of Macedonia Press, 2006.

Wallach, Wendell, and Colin Allen *Moral Machines: Teaching Robots Right from Wrong*. Oxford: Oxford University Press, 2009.

Ware, Kallistos, Bishop of Diokleia. "The Unity of the Humanperson: The Body-Soul Relationship in Orthodox Theology." *Sciences, Cutting Edge Technologies and Orthodoxy, International Scientific Conference*, October 4–8, 2000, 336–347. Holy Synod of the Church of Greece. Athens: 2002.

Warwick, Kevin. *March of the Machines: The Breakthrough in Artificial Intelligence*. Chicago: University of Illinois Press, [1997] 2004.

Wesley, Smith. "Humans Will Evolve Beyond Our Selves Via Technology? At Least the Idea Offers Some Comic Relief." *Evolutionnews,* March 3, 2015. https://evolutionnews.org/2015/03/humans _will_evo/ (03/02/2021).

Wilks, Yorick. *Artificial Intelligence: Modern Magic or Dangerous Future?* London: Icon Books, 2019.

Wynsberghe, Aimee van, and Scott Robbins. “Critiquing the Reasons for Making Artificial Moral Agents.” *Springer, Science and Engineering Ethics* 25 (June 2019): 719–735.

Xexakis, Nikolaos. *Orthodox Dogmatics C’. The Doctrine of Creation*. Athens: Ennoia, 2009.

Xionis, Nikolaos. *Man and Society, The Creation of the World and Human Nature*. Athens: Ennoia, 2020.

Yudkowsky, Eliezer. “Artificial Intelligence as a Positive and Negative Factor in Global Risk.” In *Global Catastrophic Risks,* ed. Nick Bostrom and Cirkovic M. Milan, 308–345. New York: Oxford University Press 2008. .

Yu, Han, Zhiqi Shen, Chunyan Miao, Cyril Leung, Victor Lesser, and Qiang Yang. *Building Ethics into Artificial Intelligence.* 2018. https://arxiv.org/abs/1812.02953 (17/12/2020).

Zacharopoulos, Nikolaos. “Saint Nicodemus the Hagiorite as a Teacher of Spiritual Life.” *Orthodox Spirituality, Christianity-Marxism.* Thessaloniki Theologians’ Seminar. Thessaloniki: 1968.

Zisis, Theodoros. *Man and the World under the Economy of God according to the Holy Chrysostom*. Thessaloniki: Patriarchal Foundation for Patristic Studies, 1971.

Zizioulas, John, Metropolitan of Pergamon. “Man and the Environment: An Orthodox Theological Perspective.” *Sciences, Cutting-edge Technologies and Orthodoxy*, 275–284. Holy Synod of the Church of Greece. Athens: 2002.

# Index

# About the Authors

The Very Reverend Dr. **Aristarchos-Vasileios Gkrekas** is a clergy member of the Holy Archdiocese of Athens and an associate professor at the Department of Theology, National and Kapodistrian University of Athens, where he directs a university laboratory. His academic background includes a BS in primary education; an MS in theology, conflict prevention, sustainable peace, and security; and a PhD in theology. His multifaceted work encompasses teaching, research, participation in EU programs, and leading intercultural projects. He has published eleven books and numerous articles in Greek and English journals on theological content and topics related to international relations. He has also participated as a speaker at international conferences and workshops on theology, AI, faith-based initiatives, and cultural diplomacy. Continuously engaged in international projects, he strives to enhance intercultural dialogue, peace-building, and tolerance, integrating these efforts with innovative AI applications to enrich social practices.

**Nicholas Kokkinos** is a senior digital marketing strategist with over twenty years of experience, known for implementing successful case studies and achieving phenomenal results through innovative campaigns. His master's degree in leadership, with a specialization in innovation, complements his professional skills. He is also Freeman of the City of London. Nicholas is deeply interested in the convergence of technology, religion, ethics, spirituality, and history, exploring various spiritual practices beyond Christianity, including Buddhism and Hinduism. Fluent in French, English, Greek, and Italian, he also understands other languages, showcasing his broad intellectual curiosity and understanding of global cultural dynamics. This publication marks his debut, drawing from his more than twenty-five extensive years of hands-on involvement with internet and digital technology.